Migration to the Stars

Never Again Enough People

Edward S. Gilfillan, Jr.

MIGRATION
TO THE STARS

Never Again Enough People

Robert B. Luce Co., Inc. Washington, D.C. — New York

Gilfillan, Edward S. 1906-
Migration to the Stars.

   Includes bibliographical references.
   1. Outer space—Exploration.   2. Astronautics and civilization.   3. Space ships.   I. Title.

TL793.G47   999   74-28616
ISBN 0–88331–072–4

Contents

To Elinor

with love and appreciation

Author's Preface

Any writing necessarily reflects prejudices derived from the author's background. For this reason the reader may wish to know something about mine.

At sixteen I was a chemist in a paper mill. Each day I found out what would be required of me on the morrow and each night went to the public library to find out how to do it. I was so interested in what I was doing that I skimped on high school and college. I decided then to become a consulting engineer, but soon realized that merely being able to solve technical problems was not enough — that to get good jobs one had to have credentials. Accordingly I enrolled in the Harvard Graduate School, and after four years there, chafing at the restrictive nature of the life, I got my Ph.D. in chemistry. Meanwhile I had audited courses in mathematics and physics.

With my degree I was awarded a Sheldon Fellowship for advanced study overseas, and went first to the College de France and then to the Kaiser Wilhelm Institute in Berlin, where I did experimental research in nuclear physics. This was at the beginning of the Hitler regime, and what I saw there set me thinking about social problems. Immediately on returning to the United States I joined the Naval Reserve.

At this time I found out that I did not know enough about business practices to set up on my own, and accordingly joined Arthur D. Little Incorporated, the best consulting firm of that time. After a few years there I felt that at last I knew enough to set up my own business and did so in Manchester, Massachusetts.

My first large client was the Westinghouse Electric and Manufacturing Company, where I worked at the East Pittsburgh machine shops. There I came in contact with distinguished engineers and spent many hours wandering through the great shops, watching the building of the huge machines which would supply a large fraction of the nation's power. Then the war broke out in Europe and I became consultant to the Naval Ordnance Laboratory and the Geophysical Laboratory. This was my first experience of public service and I marveled at the differences between government and industry.

When the United States entered the war I reported for active duty in the deck branch of the Navy. I enjoyed the military life, with its order and discipline. The close of the war found me in Japan where I volunteered as ship's company to take the ex-Japanese battleship Nagato to Bikini for the atomic bomb tests. When she was sunk I transferred to the technical branch and stayed in the islands to study the effects of nuclear radiation on the life there, working with biologists and coming to understand something of their point of view.

Returning to the United States I found there was a vacuum in government science. The professors who had flocked to Washington at the beginning of the war, and who had worked harder and done more there than anyone else, were tired out and returned to their universities as soon as the war was over. I moved into this vacuum and worked on the effects of atomic weapons, the techniques of amphibious operations, and the evaluation of military intelligence. While working on defenses against intercontinental missiles I had occasion to design large mirrors, and when the first Russian sputnik flew I began to wonder whether these could be used to control the weather. A preliminary calculation showed that the cost of doing this would be enormous but that it was not absolutely out of the question. Shortly after this I began to teach mechanical engineering at the Lowell Technological Institute and gave some lectures there on space mathematics. In these classes I assigned problems of space technology as exercises and my students did a great deal to clarify my thinking. Later I gave similar lectures and assigned similar problems at the

Chinese University of Hong Kong. During the same period I gave numerous public lectures on these subjects. At each there was a discussion period and from these I learned how the public was reacting to the space program. This book is a condensation of these lectures.

I have had the pleasure of discussing some aspects of this book with the Reverend Peter Lee, Director of the Tao Fong Shan Christian Study Center. I don't think he agrees with me about any of it, but I have thought carefully about his analysis of these problems and in some cases changed the text to conform to his ideas. I take this opportunity to acknowledge his help.

Introduction

We can, if we wish, move out into orbit about the earth to live and work in comfort and dignity, independently of what goes on below. We now have all the materials and technology required to do this. Not one nut or bolt remains to be designed; we shall need nothing we cannot take off the shelf.

We shall move out there, not because we want to but because we have to. There is an immediate reason for going — the earth's surface may soon become uninhabitable because of nuclear war or some other catastrophe, and we want the human race to survive — but there is a deeper and more compelling reason for going. We are what our remote ancestors were — colonists, always on the march toward better environments, always evolving, always adapting, learning how to control the physical world to our advantage. It is inconceivable that we have here and now come to the end of our long march, reduced to clinging to what we have, with no prospect for improvement — no hope. Rather we must view our present situation, with all its very real problems, as merely an overnight campsite along the way; confused, troublesome, unsatisfactory, but unimportant; an untidy place to be abandoned and forgotten. We are not near the beginning of the evolutionary process, nor near its end, but on our way to complete dominance of the physical world, to change it into something different which we cannot yet visualize.

If we add to our present technology just one new thing — thermonuclear power — we shall have the capability of moving out to the stars and colonizing them. To colonize the whole of the Milky Way will take about a million years.

This book records my search for insurmountable difficulties in the way of our long march. I have found none, and I am convinced that none exist. The difficulties I have found I have met here head on, by design and calculation, to see how they can be overcome. It is up to my readers to judge how well I have succeeded.

1.   Preview

Technology, by advancing faster than our social capabilities, confronts us with a dilemma. It offers a bewildering succession of ways we can kill ourselves off by making the earth uninhabitable.[1] Nuclear war is only one of these. But technology also offers us a way of surviving even if we are unable to control our environment and ourselves well enough to continue here indefinitely. What has been learned about space technology since the first Russian sputnik was launched in 1957 suggests that it may be possible, within fifty years or so, to establish a few thousand individuals in orbit about the earth with technology and equipment adequate to survive whatever may happen there, to increase in number, and ultimately to move out to the stars, continuing indefinitely with successive colonizations. This book explores that possibility by critical examination of the technical and social problems involved.

Even if it proves to be true, as suggested in Chapter 2, that the earth will become uninhabitable in less than two hundred years unless most of us move out into space before then, and if it is technically feasible, as outlined in Chapters 3, 4, and 5, to continue the human race indefinitely by successive colonizations, leaving behind a series of places littered with physical and human debris like abandoned picnic grounds, it does not follow that we have the moral strength to do it, or that we really want to. For to embark on such a course to perpetuate the human race would involve prodigious effort. We should have to give up some of our dearly won freedom to do as we please. We should have to submit to more bureaucracy than we have now. It would mean giving up some of our cherished social programs and abandoning philosophic concepts we have taken for granted for more than four thousand years.

Our present situation is somewhat analogous to that of men in a wrecked submarine. They have the choice of dying in relative comfort where they are or of attempting the agonizing and hazardous ascent to the surface with less than an even chance of getting there and still less hope of surviving on the surface if they make it. If some men decide to die where they are instead of facing unimaginable pain and fear I cannot fault them. They know the facts; they are making a rational decision, and I doubt that anyone is qualified to say they are wrong, but our interest inevitably focuses on those indomitable souls who elect to fight it out, and especially on the few who survive.

In discussion following lectures I have given on these problems, and in informal conversation about them, I have found people eager to go into technical details of what we might do in space and how we might live there, but reluctant to come to grips with the question of whether we should go there at all. We enjoy regulating small matters but shrink from large decisions, putting them off as long as possible, often pretending there is nothing to decide. We have successfully blanked our minds to the fact that we are living on a minute-to-minute basis, at the mercy of some unknown Russian with his finger on the nuclear trigger, going about our daily lives as though the future were a settled thing, with only a few minor problems like inflation and fuel shortages to circumvent, and with the calm certainty that we shall end our lives under much the same circumstances as we began them, with the same creature comforts and the same daily routine. That we are able so to blind ourselves to the awesome reality is a triumph of the resilience of the human mind and without this new power it seems doubtful that we could survive at all. This is a new power because it is less than a thousand years since almost everyone was deeply conscious of imminent disaster every minute of his life, with real doubt in his mind whether he could survive another day in the face of threats of starvation and disease. It has taken us more than fifty thousand years to pass from the mentality of beasts to imaginative, thinking beings, yet in less than a thousand years we have been able, first, largely to banish catastrophe from our daily lives, and second, to adjust to the prospect of imminent catastrophe without any substantial alteration of these lives. Small wonder then that we are repelled by discussion of basic questions, and annoyed when someone insists on raising them.

As privacy becomes more difficult to attain we value privacy more, and especially we value intellectual privacy. No doubt most of us have thoughts we will never mention to anyone, and views about conventions and religious beliefs which we will never divulge. There is a question of indelicacy involved; as one famous courtesan put it, it is as indecent to show what we think as what we have. Primitive and poverty-stricken people, for whom privacy is impossible, express themselves in coarse, forceful, direct terms; those of us for whom privacy is possible talk in roundabout ways designed to protect us from the impact of basic questions. A rational discussion of the space program requires us to break through this wall of intellectual privacy by raising questions we would rather not think about, and which we find disturbing.

Religion is a particularly sensitive area. Most of us are reluctant to discuss our views on religious matters, but we cannot resolve important questions without being decisively influenced by our personal views on these things which we do not mention when we discuss large issues. Thus an approach to basic problems often seems obtuse, not from obstinacy or irrationality, but from reticence; we duck these issues to avoid indecency.

In talking about these things a number of my friends have said "Well, you may be right, but I just don't want to think about it." If I offend some readers by raising issues they would rather not talk about I apologize, but I cannot find any other way to get at the questions which underlie any practical decision about how to proceed with the space program.

Others with whom I have discussed these matters react quite differently and can hardly wait to tell me where my

3

thinking is at fault. This is fine! This is what I hope my readers will do. I can learn nothing from people who agree with me.

In Chapter 3 it is found possible to establish self-sufficient colonies in space. Our experience with Skylab seems to have removed the last doubts about the feasibility of this. Such colonies will need very little from earth if they do not grow; when they grow they will need only air, seawater, granite, limestone and phosphate rock, which they will be able to get even from an uninhabitable earth. It appears that the most economical place to establish such colonies is in orbit about the earth, where apparently it will cost less than twice as much to establish a new family as it does now in the United States. Other possible locations for colonies, including the moon, the planets or their moons, or the Lagrangian points[2] of the earth-moon or earth-sun systems, appear to be more expensive to develop, but we may colonize these places when we become more affluent and can afford to do so.

It appears that the space about the earth could accommodate a population more than a hundred times as large as we have now, and we may need a population that big if we are to move out of the solar system and colonize the planets of the stars. These people would have their own agriculture and their own manufacturing and would live much as we do now.

In Chapter 4 we shall find that it is quite impossible to build ships capable of carrying us to the nearest stars and back, or to make significant use of the relativity slowdown to prolong our lives. Such ships would have to travel too fast; even if we could gather together enough fuel to drive them they would burn up on impact with the small amount of gas present in interstellar space. It is practical to build and fly slower interstellar ships capable of making one way voyages to the nearest stars. The shortest of these voyages would take fifty years or more; most would take more than a hundred years. The people who actually make the landing would be born en route in the third to tenth generation. Probably less than one in a thousand such voyages would lead to a viable colony; the rest would be lost to mechanical or human failure,

4

or would find on arrival that there was no suitable planet, or that conditions were too difficult for successful colonization. The major compromise which must be reached in designing such a ship concerns the size. The larger the ship the greater the chances of making a successful colonization; the smaller the ship the more of them could be sent out. It appears that a good compromise is an initial crew of eight.

Determination of the size of the landing party requires a similar compromise; the fewer the people the more supplies can be carried. Again it appears that the optimum size might be eight people.

It seems doubtful that a distant planet where conditions for colonization are as favorable as they would have been here a hundred thousand years ago will be found. Chapter 5 gives an hypothetical account of a landing on a planet where conditions are marginal, not bad enough to rule out the possibility of successful colonization but not good enough to make it easy.

Chapter 6 details what is known, relevant to these matters, about the cosmos. It appears likely that there are, at the present time, stars other than our sun in the Milky Way which have planets on which there is intelligent life. The question whether we ourselves may be a colony is discussed.

Chapter 7 discusses the question of men versus machines, and shows why men rather than machines will have to carry the burden of the space effort. Chapter 8 deals with the immediate future of the space program and goes into questions of levels of expense and priorities.

Chapter 9 identifies the space program as an essential part of the evolutionary process. Chapter 10 deals with our human resources and rejects two suggestions which have been made: one, that we can be changed into something different, and presumably better, by indoctrination and religious exercises, and two, that we can and should be changed into something different, and presumably better, by genetic engineering. This last could probably be done within a few hundred years, but it is difficult to see how we could get agreement as to what we should become, and equally doubtful that we could produce entities better suited to be colonists than we already are. The whole history of evolution has been the production of colonists and it seems doubtful that any planning, however brilliant, could produce individuals as well suited to be colonists as we have arrived at by the trial-and-error method of evolution.

Chapter 11 ties the preceding chapters together and contains a list of propositions which follow from what has gone before, and of whose validity the reader must be the judge.

Notes

1. "There is a serious question whether self-identification with mankind can be achieved before we destroy ourselves with the technological forces our intelligence has unleashed." Carl Sagan, *The Cosmic Connection*, New York: Doubleday, Anchor Books, 1973, p. 6.

2. No general solution has been found for the classical "problem of three bodies" concerning the orbits of three bodies — say the earth, the moon, and an artificial satellite revolving about their mutual center of gravity under their mutual gravitational attractions — but several particular solutions are known. If the orbits are all in the same plane there are certain points in this plane called the Lagrangian points, where the smallest body will tend to remain if it arrives there from the right direction at the right speed. Two of these points are on the orbit of the middlesized body and lag or lead it by 60 degrees. If, at a certain time, the smallest body is near one of these points with nearly the right speed and from nearly the right direction it will go into a stable tadpole- or horseshoe-shaped orbit about the point. An interesting discussion of this phenomenon is given in *Science News*, Volume 105, page 318.

It is entirely possible that the Lagrangian points of the earth-moon or earth-sun systems may be good places to park space dwellings, particularly since interstation busing between them may be less costly than between stations orbiting around the earth.

2.   Catastrophe

I approach this chapter with trepidation. I am in danger of convincing some readers of the exact opposite of what I myself believe. Worse, I may be caught in the toils of illogic. The trouble is that of the seven kinds of possible catastrophe I discuss here, there is only one — the terminal laboratory experiment — which I believe to be truly inevitable. Of the other six — nuclear war, pollution, overpopulation, depletion, inadvertence, and futility — we could beat any one, and possibly any two, if we could devote our entire time and strength to avoiding it. But this is just what we cannot do; we have too much else on our plate. Anyway, with seven threats, any one of them fatal if it comes up, our situation is grim. In the classic words of the GI, "Things don't look too good; let's get the hell out of here!"

It is no use talking generalities; I have to come to grips with specifics. But first there is a purely arithmetical point which must be disposed of.

Suppose you find out what a mad scientist is up to and stop him before he throws the switch. You have avoided immediate catastrophe but have not diminished the ultimate threat at all. No matter how many mad scientists you find and stop the danger remains the same. But if just one gets by you lose. This is a game you can lose but you can't win. The arithmetic defeats you.

Now let us move on to specifics.

Nuclear War:

Nuclear war is clearly possible. The bombs and the missiles exist, and there are men who know how to fire them. It is possible that negotiations like the present SALT talks will bring about a reduction in the number of missiles ready to fly. Maybe, maybe not. If there is a nuclear war it will not necessarily mean the end of the human race, though it would certainly delay everything we are trying to accomplish. What we need are soundly based probability estimates, but we do not have them. In my opinion the probability of nuclear war in any given year is now less than one percent. This figure, in my own mind, fluctuates daily as I hear the news. Purely as a numerical exercise, assume one percent. The probability that there will be a nuclear war sometime during the next sixty-nine years comes out even; if we assume that the probability of a nuclear war will remain constant for the next hundred years, the odds that there will have been one are sixty-three to thirty-seven. At two hundred years the odds are eighty-six to fourteen. But this is not the entire story. If there were an all out nuclear war at present stock-pile levels, and if this war were confined to the northern hemisphere, there would probably be more than one hundred million casualties. The culture could bear this, though somewhat slowed down, and the space program would probably not be set back more than twenty years. It would take stockpile levels several fold the present ones to set the program back a hundred years, still more to break the culture entirely, and much more to cut off the human race. Even at this level a few fish would probably survive, the radioactivity would decay to nearly nothing, evolution would resume, intelligent life would evolve, and in about a hundred million years the space program would be resumed.

This is an area in which I claim both competence and seniority. As a student I attended Madame Curie's lectures. At the Kaiser Wilhelm Institute I was acquainted with Doctors Hahn and Meitner, who later discovered nuclear fission. As National Research Fellow at the Massachusetts Institute of Technology I worked on the separation of isotopes. I was in Nagasaki a few weeks after the atomic bomb exploded there. I was present at the Bikini tests and the next year led a team of scientists back to the atoll to see what radioactivity had done to the plant and animal life there (in brief, nothing). I was then employed by the Army to work out countermeasures against extremely heavy nuclear attack.

Basic to these studies was the fact that our remote ancestors and ourselves have always been subject to nuclear radiation and have probably benefited from it. Some of this comes to us from outer space in the form of cosmic radiation greatly modified by passage through the atmosphere. Most of it comes from inside our own bodies from the radioactive element potassium, of which each of us carries about a tenth of a pound and which is essential to life of any kind for chemical reasons unrelated to its radioactivity. A small part comes from radiocarbon produced by the impact of cosmic rays on nitrogen in the high atmosphere. A substantial fraction comes from radium within the body. All foods contain measurable amounts of radium which, because of its chemical similarity to calcium, tends to accumulate in the bones. Some of the radiation entering the body from without also comes from radium contained in soil, bricks, mortar, and stone. All these radiations are known to affect genetic material, mostly destroying it, but to some extent rearranging it into new combinations. Most of these are undesirable — in fact no desirable rearrangement of genetic material has ever been observed — but it is believed that favorable mutations have and do occur, and that without them there could have been no evolution from one-celled plants to ourselves.

Soon after I got back from Bikini I was asked by the Operations Research Office of the Army to find out how many days a continuous exposure to one unit[1] of radiation would take off a man's life. There were no data on humans so I had to do the best I could from studies on animals. From these an astonishing fact emerged. Animals exposed to low levels of radiation lived significantly longer than animals not so exposed. And so that project went out the window.

Details of these experiments are of interest. The animals, mostly rats, mice and guinea pigs, were divided by lot into groups of equal numbers of individuals, experimental and control. The experimental animals were kept in cages in a lit-

9

tle amphitheater in the basement of the laboratory building. Where the stage of a real amphitheater would have been there was a hole in the floor, and deep down in this hole was a bottle of radium. The last thing the janitor did at night was to pull this bottle out and hang it above the amphitheater, and the first thing he did when he came in the morning was to put the radium back in the hole. Thus the experimental animals were irradiated all night, every night.

There is a protocol to this kind of experiment. You are not supposed to draw conclusions or publish anything until all the animals in both the experimental and control groups have died. In one experiment in which I was interested two guinea pigs in the irradiated group lived on and on after all the others in both groups had died. They held up the entire decision-making procedure. I just subtracted them out, along with two of the control animals, and drew conclusions anyway. So far as I know these two guinea pigs are still alive, happily soaking up gamma rays every night.

Trying to analyse the results from exposure of the test animals to the bomb flash at Bikini, I was puzzled to observe that more animals died on Friday than on any other day of the week. A little telephoning established the fact that this was the day the sailors cleaned the cages.

Another surprise came when I tried to analyse the statistics on smokers and non-smokers in an effort to find a norm of comparison with nuclear casualty statistics. If you choose at random a group of men, all of them fifty years old, who have smoked heavily throughout their adult lives, and compare their mortality with that of a group of non-smokers fifty years old chosen in the same random way, you find the heavy smokers live longer. An argument for smoking heavily? Not at all. The smoking helped kill off most of the weaker brothers before they reached fifty, and those who were left to be chosen in that group were tough old birds who outlasted their less colorful brothers in the non-smoking group, in spite of smoking rather than because of it. Statistics have to be used with caution, but it appears to be established that some animals, at least, live longer under lifetime exposure to up to half a unit of radiation per day. At dose rates higher than this their lifetimes fall, and at about one unit of radiation per day are the same as those of the unirradiated animals. At still higher dose rates, the average lifetime of the irradiated animals falls below that of the animals in the unirradiated control groups. Thus there is, for animals at least, a threshold dose rate below which radiation is helpful and above which it is harmful. This is an example of a general phenomenon with applications outside the problem of radiation, and in particular to the pollution problem. Copper, manganese, and zinc, considered as poisons, which they are, have dose thresholds below which they are beneficial, even necessary, and above which they may prove fatal. For some of the alkaloids, such as strychnine, the threshold is sharply defined. A certain amount helps people with heart trouble, but just a little more will kill them. Even common salt, not ordinarily thought of as poisonous, will kill at five times the optimum ration. For animals, the optimum continuously applied daily dose of nuclear radiation is about ten thousand times the pre-nuclear-age background radiation, and about a thousand fold less than the dose which is fatal in a single flash.

If you get your dose of radiation all in one flash you either live or die, and if you live you recover completely. There is a widespread belief that even if you seem to recover completely you may be more susceptible to cancer many years later, but the Japanese experience does not support this view. They had more than ten thousand people who got near-fatal single-flash doses, and for whom there is an adequate control group of similar people who got no radiation. The Japanese experience shows no statistically significant difference between the irradiated and control groups even after twenty five years.

Judging from experiments with animals, people should be able to work at reduced efficiency and to reproduce under lifelong irradiation at the rate of two units per day, and could probably survive under lifelong irradiation at the rate of five units per day. There is nothing new about working all one's life at reduced efficiency; a population in the Andes has done this for many generations, working in mines in a partially

11

numb condition. In their case, the reduced efficiency is partly due to the thinness of the air they breathe at those high altitudes and partly due to chewing coca.

As far as can be learned from nuclear war games, in which I have frequently participated, if there is a nuclear attack some areas will get much more than the average fall out and other areas much less. It is not possible to predict in advance of the attack which areas will be heavily contaminated and which lightly. It is doubtful that anyone will survive in areas of heavy fallout. In areas of light fallout restoration of farms and factories will be relatively easy. Heavily contaminated areas will probably remain unusable for a hundred years or more while lightly contaminated areas can be cleaned up and restored to unrestricted use in less than ten years.

For about three months after a nuclear attack people will have to spend most of the time in shelters. These may be surviving buildings, ruins, or specially prepared refuges. A little bit of distance between you and contamination does a lot of good. Canned or packaged food is not damaged by radiation. It is remarkable how well contaminated water is purified by stirring it up with mud and letting the mud settle. One can get by.

Radiation from nuclear fallout decreases with time. One percent more time since the explosion of the weapon corresponds to a 1.2 percent drop in the intensity of the radiation. So the problem and trouble become steadily less after the fighting stops. After about ten years only two materials, strontium 90 and plutonium 239 remain in dangerous amounts. I have not included radio cesium here because, though its half life is a hundred years, it is not dangerous because it does not tend to accumulate in the body. Strontium 90, with a half life of thirty years, is the main problem because it tends to accumulate in the bones. Plutonium 239, whose half life is five thousand years, also accumulates in the bones, where it is very harmful, but all plutonium compounds are transformed into extraordinarily insoluble materials by contact with rain or sea water, and so plutonium is rapidly and permanently removed from the biological nutrition cycles. There is no penetrating

12

radiation from plutonium 239; even next to the skin it does no harm, and is dangerous only when actually taken into the body.

The genetic effects of radiation are feared because they may be delayed, even for generations, and are not apparent at the time the damage occurs. It seems worthwhile to look at this problem in detail.

Before there was any man-made nuclear radiation, about one birth in two hundred was deformed. Among the Japanese exposed the percentage of deformed births has been the same as that of the unirradiated population within the limits of statistical reliability. In the case of animals, when both parents receive near-fatal doses of radiation just before mating, the number born is only about ten percent of the offspring to be expected from unirradiated animals, and about ten percent of the births which do occur are deformed. It is likely that there will be an increase in the number of deformed births to heavily irradiated human parents, but deformities of any kind, in humans or animals, breed out of the system rapidly, and are practically gone by the third generation, because deformed individuals are less likely to have offspring than are perfect specimens.

It appears likely that the major genetic effect of a nuclear war would be that a number of conceptions which might otherwise have occurred would not. In a world almost in a panic from the threat of overpopulation, and a world increasingly dominated by the pill, it is difficult to find any social significance or reason to worry in this possibility.

We must now consider possible very long term genetic effects. It is theoretically possible that children of irradiated parents, and their children through many generations, might be perfectly normal but still carry latent genetic defects which might suddenly appear in remote generations. When and if such latent defects become apparent, the practical result will be a decrease in fertility, though there is a remote possibility of deformed births from this cause. You hear exaggerated stories, some of them from famous scientists, about these "genetic deaths." What you are not told is that these are not

13

deaths of real people, but just conceptions which didn't happen, and that these hypothetical failures to conceive are spread through an infinite population and an infinite time.

The paradox is that if nuclear war comes soon we may weather it but if it comes later, after more stockpiling, we may not.

Pollution:

I do not think that pollution is a serious threat to the survival of the human race, or even to our capability of remaining here on the surface of the earth, except to the extent that if we were already in a bad way from other causes it just might tip the scales against us; but since so many of us are concerned with this problem I shall detail the facts as I know them.

Pollution is the addition of a noxious substance to the environment. This may be by natural causes, as when a volcano spews sulfur dioxide into the air, by indifference, as when we litter, or by necessity, as when steel mills run sulfuric acid into a stream. Pollution must be distinguished from inadvertence, as when we add what we think to be a harmless substance, like DDT, to the environment. Inadvertence may well constitute a greater danger than pollution.

The local and worldwide problems of pollution are different. I shall consider the worldwide problems first. As in the case of nuclear radiation, the problem we now call pollution is as old as the world itself, and in the worldwide case man-made pollution is just coming up to natural pollution in magnitude.

Dust

We see it in the form of a haze which becomes apparent everywhere as soon as the wind drops. As you see it from the Himalayas over India, the top of the haze is well defined and has a sharp horizon like that of the sea. Our most beautiful sunrises and sunsets come about because of haze. I had occasion to look carefully into the problem of atmospheric dust in connection with planning systems for detecting foreign atomic bomb tests. It turns out that most of the dust in the air comes from evaporation of spray from the oceans and consists mainly of common salt. Some of the rest of the dust is picked up by the wind in the dry areas of the world. A lot comes from volcanos. Some comes from outer space in the form of meteorites which volatilize in the upper atmosphere — the vapor condenses to dust. Estimates of the amount of this vary widely but it appears possible that as much as half our present soil came in this way from outer space. The amount of dust we have added from smokestacks and automobile exhausts is insignificant on the worldwide scale.

All natural and most artificial forms of atmospheric dust dissolve in the body fluids. In the course of a lifetime we breathe in pounds of the stuff, but because it dissolves it never accumulates in the lungs nor does any harm whatever. In local situations, in which man-made dust particles may be larger and composed of less soluble constituents than the natural dust, the lungs may become choked, with fatal results. On the worldwide scale, rain and snow wash out the dust in the air as fast as it gets there and a stable composition is reached. The readily soluble portions of atmospheric dust are returned to the sea and the less soluble parts are added to the soil. There is no evidence that there is more dust in the air now, worldwide, than there was a thousand years ago.

Carbon Dioxide

Perhaps carbon dioxide should not be considered a pollutant, since it is as necessary as water to plant life, but because some scientists fear that a carbon dioxide catastrophe is in the making I shall discuss it here. About 85 percent of the carbon dioxide which comes into the atmosphere is from the decay of vegetation. About ten percent comes from human activities, mostly from auto exhausts, some from home heating with fuel oil, natural gas or coal, and some from coal or oil fired power plants. But carbon dioxide is avidly captured from the air by growing plants which convert it into food and the oxygen we breathe. The turnover is remarkably rapid, plants consuming in two years as much carbon dioxide as is present in the air at any one time. This rapid turnover could lead to an unstable situation were it not for the fact that the oceans and their

15

bottoms contain more than a million times as much carbon dioxide as is present in the atmosphere and can readily absorb more or give some up to offset chance or even systematic variations in the yearly consumption by plants.

In theory, at least, carbon dioxide in the atmosphere acts something like the shutters of a Venetian blind, controlling the flow of heat from the earth to outer space but not affecting the influx of heat from the sun. If the Venetian blind got stuck — if human activities put too much carbon dioxide in the air — we might get more heat from the sun than we could get rid of, and we would all roast.[2] Measurements of the amount of carbon dioxide in the air, worldwide, and of the average temperature of the earth are difficult both to make and to interpret with the accuracy required to resolve this problem, but it would appear from such measurements as have been made that the carbon dioxide content of the atmosphere has increased slightly during the past ten years and that the earth has grown slightly cooler during the same period. Evidently forces more powerful than the carbon dioxide window are at work shaping our climate. After all, Nature arranged the Ice Age without any help from us.

Carbon Monoxide

This is the gas which kills you if you breathe too much automobile exhaust fumes. Below a certain threshold concentration, it is harmless.[3]

Carbon monoxide has been with us since the beginning of time. It is formed in the air by the oxidation of methane, commonly known as marsh gas, which comes from decaying vegetation. More than 90 percent of the carbon monoxide in the air got there that way. Since it does not accumulate there must be some mechanism which destroys it as fast as it is formed, but this has not yet been identified. Some of the carbon monoxide now in the air, worldwide, comes from automobile exhausts.

Worldwide, carbon monoxide is not a problem, but it may cause severe local difficulties in vehicular tunnels and wherever there is heavy traffic in the absence of wind.

The oil of concern in the pollution problem contains only hydrogen and carbon, with small amounts of nitrogen and sulfur. Cooking oils also contain oxygen. Of interest here are crude oil, just as it comes from the ground, and refined oils including gasoline and the fuel used in heating houses and in ships and power plants. Also considered here are natural gas and methane, which are likewise made up of only hydrogen and carbon. These substances contaminate both air and water and are worldwide as well as local problems.

But not new ones, at least on the worldwide scale. Oil has been leaking into the sea for millions of years. On the island of Trinidad there is a whole lake of residue from the evaporation of a natural oil spill the size of which boggles the imagination. The burning bush which Moses saw was probably fed by natural gas escaping from a fissure in the rock.

It has been estimated that human activities produce only two or three percent of the total oil now leaking into the sea.[4]

The sea cleans up its own contamination by converting the spilled oil into innocuous substances. First the volatile portions of the oil evaporate. Then sunlight makes the oil combine with oxygen from the air to form acidic compounds which are still brown and sticky. These acids combine with calcium hydroxide from the sea water to form hard crystalline compounds which are inoffensive and indistinguishable from sand. Thus oil on a beach is completely cleaned up in about ten years. Submerged oil cleans up less rapidly. In the Caicos Group in the Bahamas I have seen underwater stands of eel grass where the base of every blade was ringed with oil which must have been there for at least twenty years. In addition to oxidative cleanup there are bacteria which eat mineral oils much as we eat vegetable oils, and convert them into inoffensive substances.

Oil on the surface of the sea apparently harms only birds.[5] But oil also penetrates into the bulk of the sea; it, or some substance contained in it, dissolves in the water, though not completely. My son, a marine biologist, has found that blue mussels can detect this soluble portion of the oil, but its effect on them is not yet known.

Oil vapor which gets into the air combines with oxygen under the influence of sunlight to form acidic substances which are yellow solids. These constitute the "smog" now often seen over great cities on windless days. Smog is irritating to the eyes and throat. It is a purely local problem and is always the result of hydrocarbons escaping from vehicular exhausts. Natural contamination of the air by hydrocarbons, though much greater in total amount than that from vehicles, is nowhere concentrated enough to produce smog.

Radioactivity

There will have to be more nuclear power plants if only to avoid pollution from the stacks of coal and oil fired electric generating plants. These produce radioactive waste products which must somehow be disposed of, and the best place is in the depths of the sea. There has been some concern lest these contaminate the sea to a dangerous extent. All that is necessary to avoid this is to sink the stuff where the sea is more than a mile deep. Water flows very slowly at that and greater depths, taking hundreds of years to go from the arctic regions to the equator. By this time all the radioactivity except plutonium and a little radio strontium will have decayed and both of these will have become forever attached to bottom sediments. They won't be coming back.

Nuclear plants produce another environmental effect called "thermal pollution". That is, the waste heat from these plants is absorbed in river water and this may raise the temperature of the water four or five degrees F. This kills the fish and plants in the river. Sad. But as a swimmer I am in favor of warmer rivers, and if the old fish and plants are doomed I am ready to welcome the new and better fish and plants which will replace them. In the case of the sea there is no problem of thermal pollution and no hope of warmer water. I should like to have the sea five or six degrees warmer but the amount of nuclear heat in prospect for the next two or three hundred years is not enough to do it.

18

Metals

Almost all the metals are poisonous when eaten (in the form of chemical compounds of them), but apparently all of them have threshold doses below which they are actually beneficial. All of them were in the oceans before we began putting more of them there. The oceans are so large that if the metals we put in were evenly distributed throughout, none of them would be present in more than threshold amounts. But there is a catch in it. Plants need these metals for essential life processes. They sift them out of the water and hoard them. Sea creatures who eat these plants also need these metals and concentrate them still more. Carnivorous fish, including tuna and swordfish, eat these grazers, and we eat them. Swordfish and tuna caught during 1973 had nearly threshold quantities of mercury in their flesh.[6] My slide rule had said this could not happen and this makes me wonder whether other nasty surprises may be in store.

Human Waste Products

These can never be a problem on a worldwide scale. Plants — marine, fresh-water, and land — have an enormous capacity for converting these back into food and oxygen. Without enough plants to do this we should starve before being choked out by the products themselves. But human waste products can become a serious problem locally, especially in times of unrest, panic, and floods.

I do not believe that the worldwide pollution problem can in itself destroy us, but it can add appreciably to the difficulty of living and so contribute to a more complex catastrophe. Be it noted that this is a problem of aesthetics only. I would like to have cleaner water and beaches and am willing to pay my share of the cost of having them, but not on grounds of health.

The basic fact of local pollution is that it is nowhere today as bad as it has been in similar areas in the past. The smog over Los Angeles is not as dense now as it was in the late thirties. Pittsburgh was indescribably dirty then but now that people have switched from coal to fuel oil for heating houses the situation there is much improved. Gary, Indiana was a place of beauty of a clear winter morning of the early twenties, with multicolored smokes rising from hundreds of stacks, but the air was so full of a crystalline dust that you could scarcely breathe. It is a dull place now, but you breathe more easily. There is no fog anywhere now as dangerous as those in the gas-lit, coal-heated London of the late 1880's. The Merrimack River is now a stinking mess which needs cleaning up, but it is not as bad as the Kalamazoo River was in the late twenties or the Thames in the seventeen-hundreds. Local problems are still severe but the peak of the difficulty has passed.

The total area of badly polluted regions is probably a hundred times as great today as it was in 1940, but the peaks of pollution are everywhere less than they were.

I know of only three cases in modern times in which young or middle aged healthy people have been killed by pollution: one in Donora, Pennsylvania, one near Liège, Belgium, and the most recent in London. In each of these cases there occurred what is called an inversion, a weather condition in which warm air overlies cold, and in which the fumes from automobile exhausts and chimneys cannot rise and so remain in the immediate area. This is a common condition in the northern United States and I have seen it many times early on winter mornings at Lowell, Massachusetts. This condition is now well understood and forecasted. Local officials will recognise the danger before it becomes acute and take steps to prevent casualties. There is little likelihood that this condition will ever again cause the death of a healthy person.

Less than a thousand people altogether died in these three instances. Local air pollution by volcanos killed about sixteen thousand people in Pompeii in 79 A.D. and thirty thousand in St. Pierre in 1902.

Modern water purification plants are effective in removing disease-carrying bacteria, but add chemicals, including chlorine gas, which spoil the taste of the water. It has been claimed that the action of the chlorine on carbon compounds present in the water produces substances which cause cancer, but it is too early to be sure this is correct.

Human wastes are a danger only under panic conditions. I lived for more than a year in China in an area where these wastes were spread on the fields. The stench was all pervasive, and we had to boil all water and eat only well cooked food while it was still hot (flies will not light on hot food), but this was entirely practical and I had no digestive disturbances while I was there. I am told that there are parts of central Europe where these same customs prevail.

Trash and garbage disposal are vexatious problems but they are well within the capabilities of local authorities. Abandoned automobiles will not be a problem much longer as iron moves toward the ranks of precious metals. It will soon be possible again to sell an old car rather than abandon it. Plastic containers give trouble because apparently they will last for hundreds of years in the sun and air. The beaches of the world are cluttered with them now, but these can be gathered up at small cost and on some beaches this is done. Though it is not generally realized, plastics will burn, though not as easily as wood and paper. Chlorinated plastics of the saran type give off poisonous fumes when burned.

There is at least one case in which anxiety over local pollution has led to unnecessary expense and some harm. Some of our New England towns are spending money needlessly and depriving their soils of needed materials for fear of dust in the form of smoke, and have prohibited the burning of wood and leaves. Wood smoke contains nothing which does not readily dissolve in the lungs, and is of no possible danger to prudent people. It may, of course, be necessary to limit the production of wood smoke where it blows across highways and reduces visibility there. Burying wood and leaves, as we do now, besides being expensive, deprives our soil of the potassium it needs and normally gets from smoke and ashes.

Inadvertence:

Actions taken before we know enough about what we are doing could endanger our survival. The use of DDT is an example. Apparently this rather inert chemical lasts almost forever, going round and round the food chains, particularly through insects and birds. The danger from this particular substance does not seem to be very great but the fact that it was used in large amounts without realization of possible danger is a shocker. Insecticides of the fluorophosphate type were available before DDT came into use. They are more effective than DDT, are rapidly destroyed by air and water, and do not enter the food chains. They are not used because, being highly toxic to humans, they are difficult to apply; but they could replace DDT.

It has been suggested that the aerosols[7] we now use may become a danger to us twenty or thirty years from now, and that if we do not discontinue use now it may soon be too late. Many aerosols are based on fluorocarbons and these are so inert that they are not destroyed by air and sunlight in the lower atmosphere and thus, many years later, may diffuse into the upper atmosphere where they may destroy the blanket of ozone which protects us from the ultraviolet light of the sun. There are only a few hundred pounds of ozone up there. In my opinion, the rate at which ozone is being destroyed by the very act of shielding us implies an equally fast rate of production, compared to which the rate of destruction by fluorocarbons may well be negligible, but I cannot be sure. The chances that we are endangered by this particular form of inadvertence are probably less than one percent, but there are hundreds of such potential dangers, some of which we don't yet dream of. One of these may get us yet.

Many drugs and food preservatives introduced during the past twenty years carried with them dangers then unsuspected. Thalidomide was an example — it led to hundreds of deformed births. Mercury poisoning, which recently led to the death of a number of Japanese who had eaten locally caught tuna, also came as a shock. The danger of mercury poisoning had been appreciated for many years, but the idea that enough mercury to be dangerous could accumulate in a single fish seemed arithmetically so small as to be ridiculous. But it happened. Most dangers of this kind will be recognized and avoided before there is any real problem, but just one sleeper, taken into the body and retained there for twenty years or

22

more without any sign of trouble, might yet be the end of the human race.

Biological agents — bacteria, viruses, and enzymes — now being used in increasing amounts in industry constitute, in my opinion, an even greater danger. Bacteria are even being used to clean up oil slicks at sea, as previously noted. They are also used to ferment grains into chemicals, as in the production of acetone. These materials not only have the capability of growing and reproducing from the everywhere plentiful proteins but also the capability of changing abruptly into different forms which are also capable of growing and reproducing. Suppose, as might happen, that one of these agents metamorphosed into a deadly contagious disease with an incubation period of a month or more.[8] By the time it was found it would have spread all over the world and there would be no vaccines or proved treatment for it.

Similar things have happened in the past, though not from man-made biological materials. There was the Black Death in England and Europe, syphilis among the American Indians, and measles among the Eskimos.

There have been minor crises from the introduction of the wrong animals and plants in new places. Rabbits were taken to Australia as food animals but were soon eating up the crops. Mongooses were brought to Puerto Rico to kill off the snakes, which they did, and then they went to work on the birds. The water hyacinth was brought from Africa to America because it was pretty; now it chokes many of our southern streams.

The problem is even older. When the Romans began to use lead pipes and lead cooking pans they had no idea that these might be dangerous and they never found out that they were. They also had bronze utensils but did not use these for cooking because they knew how easily copper got into food cooked in them and how poisonous copper was. They knew these things because the effects of copper poisoning are immediate and because contamination by copper stains foods blue. The poisonous effects of lead are delayed and not easily identifiable as such, and contamination by lead does not betray itself by color. The Romans used lead pans — more convenient than the clay alternative — without any thought of danger. But the lead which is found in the bones from Roman tombs shows the danger was there. My uncle, Doctor S. Colum Gilfillan, has spent half a lifetime studying this problem. He concludes that lead poisoning played a significant part in the fall of the Roman empire.[9] Old Bostonians may have been on the same road out. It has just been found that the water to the houses on Beacon Street comes through lead pipes and is significantly contaminated thereby.

Depletion:

Our reserves of high-grade iron ores are gone. High-grade copper ores are being rapidly exhausted. The known oil reserves in the United States would last only about ten years more if we didn't import oil. Some people read doom into facts like these. We would indeed be doomed if we tried to go on forever using the same raw materials for the same purposes as we do now but, quite apart from the question of depletion, no one has any idea of doing this. Aluminum and magnesium are replacing iron in automobiles, structures, and machinery, not just because iron is becoming scarce but because, for the same strength, aluminum and magnesium are lighter. A Volkswagen contains about twenty five pounds of parts made of magnesium which were formerly made of steel. The visible supplies of aluminum and magnesium exceed by many fold the probable requirements for the next two hundred years. Although it does not perform as well and is difficult to use, aluminum is replacing copper in power lines and electric wiring because it is cheaper. Aluminum itself will, in some applications, be replaced by sodium, which is a better conductor of electricity and is cheaper still. Plastics, which can be made from the very large supplies of limestone and coral available, are rapidly replacing metals in structural parts and even in such exacting applications as the blades in the compressor stages of jet motors. Oil and coal are being replaced by nuclear sources of heat. As coal and oil become scarce they will be used mainly as raw material in the chemical industry and even there will ultimately be replaced by carbon from limestone. I find nothing to alarm me in our present supply situation.

Let us look a little further. The space program, as I visualize it here, will require raw materials in amounts greatly in excess of what we now use. The only raw materials available in sufficient quantity to support it are air, sea water, limestone, granite, and phosphate rock. But these are sufficient! They contain everything necessary to support life in space and to build there first, accommodations, and then interstellar ships. Granite will be easier to separate into oxygen, silica, aluminum, iron, uranium, and other metals than it is here on the ground. Down here we have to melt it in some kind of crucible, but at its melting point granite dissolves everything but platinum that we could make a crucible of. In space no crucible will be needed. In the absence of gravity the granite, heated by induced electric currents, will collect into a ball, and the materials we shall need are distilled out of it one by one. The high vacuum needed for the distillation comes free in space.

I foresee no danger to the human race from depletion except possibly as an aggravating factor in an already desperate situation.

Overpopulation:

Strictly speaking, overpopulation is impossible. As you get more and more people each gets less and less food until some people get too weak to reproduce and the population levels off. This has happened at least twice in India and at least once in Szechuan Province in China. What I mean here by overpopulation is a situation in which there are so many people, so much noise, so little space, and so little privacy that it is not worthwhile to go on struggling.

The bearable limit on population is probably less than that imposed by food shortage. I have lived in China in Szechuan Province, up above the Yangtse Gorges, where there are six thousand people per square mile, where food supply was the limiting factor, and crowding such that no bed was ever unoccupied, people sleeping in relays, yet they remained good-natured, apparently calm and contented. I have also lived in Hong Kong, where space, not food, is the limiting factor. There, in some resettlement estates there are more than fifty thousand people per square mile living in high rise apartments, as many as five people to a ten by twelve room. Here too the people are good natured and tolerant, though noisy and aggressive. I have not yet seen a condition where overcrowding is becoming intolerable. The nearest thing to crisis is perhaps the overcrowded beaches of France and Spain, or the traffic jams in Europe. The evidence is contradictory but apparently we have a long way to go before overcrowding becomes a serious menace.

I am aware, of course, that millions of people are now dying of starvation. I am equally aware that millions of people were dying of starvation at times when the population of the world was less than half what it is now. Starvation is a political, not an economic problem. The present starvation is not caused by a worldwide food shortage or by worldwide overpopulation, but by local political situations. This chapter is addressed to the narrow question of whether the human race can survive here. As I have seen with my own eyes in India, starving people are not a menace to anyone. It is people who, within two decades, have greatly improved their standard of living, and who demand still greater improvement forthwith, who may bring us down.

We shall not suffocate in our own excrement or automobile exhausts as is commonly supposed. Plants still cycle more than a thousand times as much carbon, nitrogen, and sulfur, as we do, and if we shift to the gas turbine instead of piston types of internal combustion engines, and to low-sulfur fuels (perhaps liquid hydrogen) automobile exhausts will become quite innocuous. The forests and sea have an enormous capacity to clean up human and mechanical wastes.

The population of the United States is increasing but during the last twenty years both the area of land in cultivation and the number of people in agriculture have decreased. Here is our hope to continue to draw ahead of the Russians. The strength of a nation is measured by how few rather than how many people there are on farms. Six percent of our people feed the other 94 percent, while in Russia 30 percent feed the other 70 percent. By pumping the Mississippi River back over the Rockies — a project quite feasible with nuclear power — the New Mexico desert could be made to produce as much food as we are now eating, but such an undertaking would be a mistake. By using nuclear power to de-salt water the Mongolian, Australian, and African deserts could be made to produce more food than is now being consumed in the entire world, but this would not diminish the number of people who are starving. As mentioned above, the reasons for starvation are almost entirely political and are only indirectly related to food supply. I was in Calcutta at a time when thousands were dying of starvation there every day, yet in the botanical gardens a few miles out of the city hundreds of tons of rice were rotting in bags right out in the open. Even so, I shall make a computation of the limits which food supply and population could approach if the political problems could be solved and if we could all stand each other at very close range.

The calculation is not as simple as at first appears. About 15 percent of the world's population, most of them in western Europe and North America, eat substantially more than is good for them. About 70 percent, mostly in eastern Europe and northern Asia, eat enough food to keep them in good health, but no more. About 15 percent, mostly in India and Africa, get so little to eat that their health and life expectancy are impaired. Most of us overfed eat red meat. To produce a pound of red meat an animal has to eat about ten times as much vegetable material (mostly grass and corn). We could get by on perhaps two pounds of fruit and grain and keep in good health if we were vegetarians — which the Lord forbid! A piece of fish on our table corresponds to more than a hundred times its weight in marine plants which we could eat if we had to. A very rough calculation based on this kind of figures shows that if we ourselves ate all the plants we could possibly grow on land and sea, instead of processing most of them

27

through animals, the earth could support a population more than a hundred times what we have now.[10]

If the land areas of the earth were watered, fertilized, and intensively farmed they could produce more than ten times the food they do now. The oceans, if intensively cultivated, could produce about ten times as much food as the land areas. One would merely have to add nitrogen compounds, manufactured from air, water, and limestone, and much smaller amounts of phosphorus and silicon compounds, to make the whole surface of the sea a mush of edible plants. These would look and taste about like baby food.

The world population is now increasing at such a rate as to double about every forty years. The time to double[11] varies widely from country to country but it happens to be about forty years in the United States at the present time. If this rate remains constant, worldwide, it will be more than 200 years before the population of the world is necessarily limited by shortage of food. But a doubling time of as little as five years is biologically possible and has been reached on occasion in some provinces of China, and so overpopulation could be upon us in as little as fifty years.

Many people consider population control to be the most urgent problem before the human race, and in the short term they may be right. Certainly I would rather not have the earth any more crowded than it is now. But if one accepts, as I do, the postulate that the destiny of the human race is to colonize the stars, there can never again be enough people. It will take a population at least ten times our present one to support a program of building and launching interstellar ships. Most of these people will be living in space, and the population of the earth — by then a vast quarry — will perhaps be less than it is now. That does not mean that we should have fewer births here. When the space program really gets under way perhaps four out of five children born here will be sent into space before they are six weeks old, there to become not astronauts or scientists but construction workers.

I do not think that overpopulation itself will bring us to catastrophe but it could become a contributing factor.

Futility:

Mass insanity due to the futility of much of contemporary life may yet be the end of us all. So many of us spend so much of our lives doing things we would rather not do in ways we particularly dislike that we are led, often quite suddenly, to wonder what it is all about and why we put up with it — and it is hard to find convincing answers. The man in his thirties who during his working day endures a capricious and arbitrary boss, who has a nerve wracking trip home on the throughway to a nagging wife and children who scream at him, spends his evening trying to answer an obscurely worded note from the Internal Revenue Service, has difficulty getting to sleep, wakes to a nagging wife and screaming children, and then dares the freeway back to the hated boss, may not be altogether satisfied with the thought that he is helping to build the economy of the greatest nation on earth and raising a fine set of children to carry on the holy tradition of patriotism and service. He may get the feeling that he is being cheated and that there must be a better way to live. Sometimes he bolts, and this is a healthy stabilizing phenomenon — other bosses, wives, and children take note and refrain from bearing down quite so hard on the old man. Even if many bolt there is still no trouble, but trouble is on the horizon. One raindrop drifting down through the air produces nothing more than a splash when it hits. Many raindrops together, so long as they drift through the air rather than with it, produce nothing more than a rain squall. But as the number of raindrops increases a crisis ensues, sudden and catastrophic. Instead of drops drifting through the air, drops and air fall together, not at thirty miles an hour but at one thousand miles an hour, and we have a cloudburst. In the New Mexico desert I have seen a corner of a cloud break off and fall to the ground in seconds. At Bikini I watched an atomic explosion toss an incredible number of drops of water a mile high. Air and water fell back together, exactly as solid water does in a plunging breaker, and those of us watching thought we saw a wave a thousand feet high rushing toward us. This was the famous "base surge". If many people rebel in dramatic fash-

29

ion against the futility of their lives this is healthy and diminishes the frustration of others, but if too many bolt at the same time the different phenomenon of mass insanity ensues.[12]

Such phenomena are known in the animal kingdom. When an area becomes too crowded with lemmings they scatter in all directions and some go into the sea. Occasionally schools of whales run aground and perish, perhaps because they are tired of being whales. One can find historical examples of mass insanity. The First Crusade was touched with it, as were the disturbances in Europe in 1848. There is the case of the Marie Celeste in which the passengers and crew of a ship all went overboard in calm weather in a crisis which appears to have lasted less than ten minutes. Ditto a group of naval aviators who all went into the sea together in 1922 for no apparent reason. Mass insanity is an ever present danger of possibly fatal proportions.

To have stability — to be free of the danger of mass insanity — it is necessary for a substantial fraction of the population to understand what the leadership is trying to accomplish. It is not necessary for the population to approve these aims. It is doubtful that there has ever been a time in history when as much as one percent of the population wanted to do what the leadership made them do, but in spite of this, or perhaps because of it, we have come from caves to sputniks. Probably few people in the communist nations really want to sacrifice the comforts and plenty they could have to do away with people who are better off than they are. They don't like the goal but they understand it and can live with it. Not so our present liberal arts student. He suspects that much that he is made to study and regurgitate is nonsense — that the professors who torment him are something less than great minds — and he is right. He suspects that the world into which he is about to move neither wants nor needs him — and again he is right. He has been taught to question everything he hears but has not been taught how to find answers to his questions. Small wonder that he turns to drugs or violent behavior. I believe that most of these students would accept almost any plausible goal — even the preposterous idea that they should devote their entire lives to getting other people to the stars — provided their contemporaries accepted it too. They would, I think, prefer to lead drab, regimented lives devoted to a clearly formulated goal — any goal — which they could actually reach, than to pursue a utopian goal which they know, deep down inside themselves, is a philosophical illusion and forever beyond their reach. The concept of a social order in which everyone is sincere and honest, where people are kind to each other, where there is justice for everyone, and where everyone lives with dignity as a respected and valued member of society, sounds good, but it is a mirage, nothing more, and the pursuit of it leads to mass insanity.

Futility is destructive in many ways. It is unlikely that we shall all, for any reason, rush headlong into the sea, but we might on a certain date decide that the only worthwhile thing to do was to grow the poppy. Or we might set upon each other with clubs, each certain that everyone else was secretly tormenting him. Or we might just sit down resigned to starving to death as I have seen people do in Calcutta. I think that futility constitutes a greater danger to us than war, pollution, inadvertence, depletion, or overpopulation.

Laboratories:

The mad scientist used to be a comic figure. He isn't any more. He may be the death of all of us.

Catastrophe by laboratory experiment could be either deliberate or inadvertent. A scientist may think himself into a sense of mission in terms of which the fate of the human race seems unimportant, and carry through an experiment which may make the earth uninhabitable because that is, in his judgment, the right thing to do. Or he may do something, knowing it may destroy the earth, deliberately taking that chance without consulting the rest of us, feeling that the completion of his mission is worth the risk involved. Or he may simply overlook the deadly consequences of what he is about to do. In any case, he does it, and we have had it.

31

The problem of the deadly laboratory experiment has two facets — the physical possibilities for destruction and the nature of the experimenter.

Historically great scientists have been intensely dedicated men[13] and women. Madame Curie made her crucial experiments in a straw-thatched shed in freezing weather. She had little equipment or money, but nothing could have stopped her. She married, it is true, and had children, but her husband was a scientist as dedicated as herself. All that mattered to either of them was their work. An ordinary person may occasionally enjoy sitting on the beach and watching the girls; he may also be a competent and productive scientist but never a great one. Almost all the present day productive scientists are, outside the laboratory, just like the rest of us, with all our diverse family, social, political, and sports interests. These are not a danger to us. But dedication corrupts as surely as power, and absolute dedication corrupts absolutely. All sense of proportion is lost, and the mission becomes the only thing of consequence. A truly dedicated man or woman may destroy us yet.

This would be of no importance if the physical possibilities for destruction did not exist. Probably they did not prior to 1920. Michael Faraday was dedicated enough but quite harmless because almost everything he used in the laboratory he had to make with his own hands, and he was thus safely limited as to what he could do. A modern scientist does not make much with his own hands; he buys things and puts them together and is limited only by the amount of money he controls. Still, it is physically difficult to accumulate enough of anything in one location — explosives, nuclear materials, chemicals, or biological agents — to destroy the world. All that a laboratory run by the maddest scientist can do is to trigger off a catastrophe which was always possible in any case.

One such possibility is the formation of helium from the hydrogen in the sea. Vast quantities of heat would be liberated in this process — enough to vaporize the sea and bring the land temperature everywhere up to more than 1,000° F. There are two kinds of hydrogen in the sea — one heavy (usually called deuterium) and one light. The heavy kind combines to form helium more readily than the light but there is less of it, about one in ten thousand of all the hydrogen atoms in the sea. This is the fuel which will get us to the stars. The more abundant light hydrogen does not transform into helium so readily, but in the sun it does so and this is the source of the sun's heat.

Once started, the conversion of most of the hydrogen in the sea into helium would take only a few seconds. It would be like a gigantic flash spreading instantly around the world. That such a catastrophe is difficult to initiate is proved by the fact that it has not happened during the three billion odd years the sea has been in existence. According to present day nuclear theory, you would have to heat more than ten thousand cubic feet of water to more than a hundred million degrees F. to start it, but once started it would be uncontrollable. A laboratory experiment which would heat more than that much water that hot is physically possible. Whether it would ignite the sea no one can know in advance.

It nearly happened. When the second atomic bomb was exploded in the sea at Bikini it heated a lot of water very hot. Many people at Bikini and all over the world were worried over the possibility that the sea might ignite. Even I, just a deck officer, knew what might happen, and in the last few seconds before this bomb went off I was sick with fear. I was not in the flagship but I am told they had a wild night there, just before the test, with a dozen eminent scientists begging the Admiral to stop the test for fear of destroying the world. I have seen a file of letters to President Truman from distinguished scientists of several nations imploring him to stop the test. But neither the Admiral nor the President would call it off.

Admiral W.H.P. Blandy, commanding, was as solid and responsible a citizen as this country will ever produce, but he was also dedicated to the good of the United States Navy. He was unquestionably aware that a risk to the entire world was involved. It is easy to follow his reasoning. If the danger was real the Navy should have recognized that fact a year before the test, not a day before. If he called the test off the Navy would look pretty silly. If he went ahead and the sea did not ignite, fine — and if it did the Navy would have no problems, nor would anyone else.

A similar situation exists with respect to the air. Nitrogen and oxygen, if heated hot enough in sufficient volume, can combine to form silicon plus other products. The reaction would liberate enough heat to raise the temperature of the air millions of degrees, and as in the case of the ocean, the flash, once started, would go right around the world. There was a distinct possibility that the first hydrogen bomb exploded in the air would light it off. Since the air has not ignited it is improbable that the necessary combination of large volume and high temperature can be reached by hydrogen bombs, but other types of nuclear reactions, more powerful than the combination of hydrogen to form helium, are possible and are already being carried out on a small scale in laboratories. If these experiments continue on larger and larger scales, it will be only a question of time until the atmosphere ignites.

Every year vast explosions are seen in the sky. One of these, in 1054, illuminated the night sky to almost daylight level. The best scientific evidence is that these explosions are caused by nuclear reactions similar to those described here. Probably most, if not all, of them come about naturally. A star shrinks too rapidly and is thereby heated to a point where new and explosive nuclear reactions begin. It is expected that, purely from natural causes and as part of the orderly evolution of a typical star, in about five billion years our sun will swell up to so large a size that it will engulf the earth. It is impossible to tell at the present time whether these explosions begin in stars or on associated and still invisible planets. If the latter, possibly some intelligent race had a scientist too powerful and too mad.

Biological experiments appear quite as dangerous[14] as nuclear, and could be made with smaller budgets.

As medical knowledge and facilities become more extensive we lose more and more of our natural defenses. This is of no consequence as long as we deal with known diseases, but

34

renders us more vulnerable to new types which might be produced in the laboratory purposely or inadvertently. As stated earlier, the condition for catastrophe is that such a new disease has a long incubation period so that it can spread over the entire world before the danger from it is recognized and an adequate treatment developed. NASA recognized this danger when it brought the first men back from the moon. Though the possibility was remote, they might have acquired some new disease there. They were quarantined on return.

There are genetic dangers as well. The code which determines what we are is being broken and when it is, genetic material will be synthesized. Such material can reproduce itself without limit and quite unaided. The possibilities defy the imagination. There is a similar danger in the field of machine design. From machines which repair themselves to machines which design themselves; from machines which have intelligence designed into them to machines which ponder whether we are really necessary; these are steps which may be taken.

Mind distorting drugs are another danger. The LSD panic was a closer run thing than is commonly realized. A moderate sized laboratory could easily produce enough LSD to madden the whole world. A new drug so pleasant and addictive that no one can resist it is a distinct possibility. This danger is increased by the sense of futility which now grips a substantial fraction of the world's population. If society does not find a way out of futility, drugs will. They already do.

Perhaps we could avoid all this trouble by prohibiting further scientific research. I have combed through history and cannot find a single case of a prohibition being completely successful. The scale of whatever is being prohibited may be reduced for a while but it always springs back. We have banned books and burned books but people still read. In communist countries people are forbidden certain thoughts, but no one knows how many people still think them, perhaps oftener because of the prohibition. We forbid the use of marihuana in the United States today quite without effect. Prohibition brought us worse liquor than we had before. If we try to prohibit research, or even to limit it in any significant way,

35

the clandestine research which will follow will be more dangerous than what we have now.

The reader will have noticed that while I started out to show why the earth is unlikely to continue habitable for much more than two hundred years, I have gone a long way toward proving the opposite. I am inclined to believe that except for the fatal laboratory experiment any of the other threats could be parried if it were the only threat. The trouble is that there are too many threats, that we are already somewhat tired, and that if we spend our energy coping with one threat we may fall victim to another. All in all, I still doubt that we have much more than two hundred years to provide a way out.[15]

Notes

1. The most commonly used unit of radiation dose, and the one used here, is the Roentgen, after Konrad Roentgen, who first worked extensively with X-rays, and who was the first to be hurt by them. It is defined as that amount of X-radiation which produces one electrostatic unit of charge per cubic centimeter of air at standard temperature and pressure. This definition is somewhat ambiguous and difficult to apply to absorption of radiation by liquids and solids, for which purpose it is conveniently redefined as that amount of X or gamma radiation which, at the surface of the solid or liquid on which it impinges, deposits energy in the amount of 82 ergs per gram. Absorption of one Roentgen of radiation raises the temperature of water .0000032 degrees F.

The background radiation to which we are all exposed is about .0000010 Roentgens per day. The maximum amount of radiation which industrial workers may receive under AEC regulations is .043 Roentgens per day, but some workers, in practice, get many times that. Single exposures to X-rays for medical purposes range from one to ten Roentgens. The minimum single dose whose effects can be detected clinically is about ten Roentgens. The fatal dose, if received all at once, is between 400 and 600 Roentgens. Larger doses can be tolerated if received over a period of time. I have interviewed one of Roentgen's laboratory assistants who received more than 10,000 Roentgens during his lifetime. He was scarred all over but was otherwise in good physical condition, and had children and grandchildren.

The single dose required to knock a man out in seconds is probably over 10,000 Roentgens.

2. More on this in Appendix 4.

3. This threshold may vary by race or habits of life. In China, in cold weather, it is the custom to supply some heat indoors from trays of sand on which a few lumps of charcoal glow. This looked bad to me. I was sure some carbon monoxide was being produced, but figured that if the Chinese could take it I could. I was wrong.

When you take a hot bath in a room where the temperature is near freezing the room fills with steam and you can't see much. While I was luxuriating in such a bath the servants slipped in several trays of glowing charcoal to warm me. I did not see these. Suddenly I became aware of a deafening ringing in my ears and a feeling of impending disaster. I barely made it out of there with my life, to lie gasping in the snow.

Whether there really is a difference among races in their tolerance for carbon monoxide I do not know.

4. *Chemical and Engineering News*. December 13, 1971, p. 35.

5. What seems to be an oil spill washed up on the beach may, in fact, not be mineral oil and be due to wholly natural causes. Late one summer the sea off Manchester, Massachusetts, was grey with a bloom of pollen from seaweed. The next day was stormy with heavy surf. The pollen broke down, liberating its contained oil. Ordinarily this would have calcified and sunk to the bottom unnoticed, but in the surf the calcified oil was churned into foam which piled up on the beaches. There the receding tide left it and sunlight did its work; when the tide came back there was a floating sticky mess. Some vexed cormorants were thrashing around in it, unable to take off. Everyone thought it was an oil spill and speculated where the oil might have come from.

6. It was thought then, on the basis of analyses of samples of fish caught scores of years ago, that this was no new thing, but Gibbs, Jarosewich, and Windon, in their paper entitled "Heavy Metal Concentrations in Museum Fish Specimens: Effects of Preservatives and Time" published in *Science*, Volume 184 (April 26, 1974) doubt that the analyses of old samples of fish are significant. The question remains open.

7. *Chemical and Engineering News*. September, 1974, p. 27.

8. The incubation period of leprosy is about ten years.

9. S. Colum Gilfillan, "The Inventive Lag in Classical Mediterranean Society," *Technology and Culture*, Vol. III, No. 1, 1962, pp. 85-87.

10. The concept of adequate nutrition is complex. In China I had occasion to compare my own health, derived from a diet of red meat and much exercise, with that of coolies who ate mostly rice with a trace of salt, a trace of fat, and a few scraps of fish or chicken. I could have gone about twenty days without food and still been able to get around, and to survive. They would die in five or six days without food. I could hike for two or three days without food and hardly miss it; in three or four hours without food they would run down like clocks and be practically helpless until stoked with three or four bowls of rice. But they could, and did, trot along for miles carrying loads of merchandise which I could not even lift off the ground. Who was undernourished?

In China I often saw pigs given shots of opium, slung between poles, and carried to market by four men, who wanted for themselves the fat the pigs would burn up if they were allowed to walk to market. There is not much animal fat in the Chinese diet, but what little there is, is essential.

11. To compute the time to double from the average family size, subtract 2.3 from the average number of children per family in the country considered and divide the remainder into 60. Note that the time to double may be negative, as it has been historically in some countries.

12. The relevance of this analogy has been questioned, and because I think it goes to the crux of the matter I shall elaborate. The point is that as many people become dissatisfied with the futility of the lives they are leading this will produce, at first, no change in the social order. As the number of these people increases there will come a point, sudden and catastrophic, at which the entire social order changes. The change is abrupt rather than gradual. The French Revolution is an example.

Take ice at a temperature of few degrees below 32° Fahrenheit. As you raise the temperature of the ice this at first produces little effect; the ice becomes a little softer and a little bulkier, but it is still ice. As you continue to warm it the ice gets to 32° F. and then it melts — all of it — and you have a substance, liquid water, of completely different properties. The phenomenon of sudden and devastating social change has much in common with the melting of crystalline materials.

13. There has been one exception, Lord Rayleigh, who was the most lucid and productive scientist of his time. He wrote about all the technical problems of his time, cutting through detail and getting to the heart of the matter so brilliantly that even now, before I start a new project, I look first to see what Lord Rayleigh said about it. He was marvelously productive in the laboratory too, but as a gentleman he had his servant actually do the experiments under his close observation. Outside of the laboratory Lord Rayleigh was a perfectly ordinary English gentleman, and does not appear to have been dedicated.

14. Apparently this view is shared. As reported in *Chemical and Engineering News* for July 29, 1974, p. 4, a group of research scientists in concert with the National Research Council has urged a moratorium on genetic manipulation of micro-organisms for fear that highly dangerous types may be unintentionally produced. The National Institutes of Health is issuing guidelines to its contractors doing genetic hybridization indicating precautions to be taken in their work.

15. As reported in *Chemical and Engineering News,* March 16, 1972, p. 2., Dr. Dennis L. Meadows of MIT forecasts catastrophe about the year 2050.

3.   Living in Orbit

After the space program reaches a certain magnitude, it will be possible for men to live and increase in number in space, dependent on earth or some other planet or moon only for air, water, limestone, granite, and phosphate rock, and for only limited quantities of these. Our space oriented descendants can live comfortably out there even if the surface of the earth becomes so badly contaminated that no one can live on it. The earth will be simply an uninhabited quarry. But much of the grand scheme of living out there, with many of its component parts, is untried. How then can we be sure that the structures and procedures visualized here will work? We can't be sure. All that I or anyone else can do at the present time is to design every possibly significant detail of such structures on the basis of present scientific and technical knowledge, using only materials and devices which can be bought now in the open market, and see whether any numerical difficulties present themselves. This I have done with the aid of many engineering and science students at the Lowell Technological Institute and the Chinese University of Hong Kong. We found that all necessary structures and apparatus could be built from materials now available in industry and fabricated by techniques now in use. As the designs progressed the numbers — dimensions, weights, temperatures, stresses, and so forth — came out so easily and naturally as to suggest a certain inevitability. Every practicing engineer knows this feeling — when the numbers fall into place almost of themselves he can be sure that he has finally got it right.

Some of our initial ideas ran into numerical difficulties

and had to be abandoned and replaced by practical solutions to the problems of living in space. For instance, when you start to think about living on the moon you notice that day and night there are each fourteen days long. It gets too hot during the day and too cold at night. One way to fix this would be to speed up the rotation of the moon about thirty fold, when day and night would be of about the same length as they are here. In principle this could be done by attaching rockets to the lunar surface near the equator, aiming them a little above the lunar horizon, and firing for a long enough time. The impracticality of this scheme became apparent when we calculated the total power required which turned out to be more than a thousand fold what we could hope to generate. So we gave up this idea and turned to devices in orbit about the moon — half mirrors and half Venetian blinds — which would alternately reflect sunlight to and away from the moon, thus providing day and night at any desired interval. This time the weight, cost, and power turned out to be practical.

The numerical criteria for accepting or rejecting an engineering design are somewhat arbitrary, and there is an element of comedy in the way they are reached. I cite the following example of the way it is sometimes done. I once served on a committee to devise exotic[1] means for shooting down incoming intercontinental missiles and to test these numerically. After grave deliberation it was decided that no weapon of which a single unit would require more power than was then being generated in the United States could be considered practical.

The working engineer discards more designs than he completes, almost always on numerical grounds. In the beginning design is more related to art than to science. He draws it as he intuitively feels it should be. Then, working from preliminary drawings, he calculates weights, stresses, temperatures and costs. If any of these come out unreasonably high or low he alters the design accordingly. Sometimes this can be done effectively — more often not. Usually when you encounter a numerical difficulty you have to begin all over again with a different concept. A successful engineer learns early not to fight facts but to ride with them. When he finds himself in numerical trouble he cuts his losses, without regret and without dismay.

If as he examines a preliminary design, he finds his weights large and his temperature differences small, he begins to wonder. He suspects that he is designing in more material than he needs to, that he is not working his materials hard enough, and that there must be a simpler, less expensive way to do it. You learn to sense and be wary of any number which seems to be just a little out of line. You also learn to sense when a design has jelled — when the numbers computed one by one fall into a sensible pattern — and you know the design will work. There are very few failures in practical engineering. Only those who develop early a sixth sense of the fitness of things survive beyond subordinate positions.

Designing equipment for use in space presents no special difficulties if the product is to be built and used within a few years, but designs whose purpose is merely to test feasibility, such as those considered here, immediately run into a paradox. You must assume that the materials and know how now available will be used. When you run into numerical difficulties you are tempted to assume that by the time the structure is actually built there will be stronger metals, plastics which will stand higher temperatures, better chemical processes for processing raw materials, and so on. This assumption is correct. There will be, but you cannot design them in. To do so would be to take leave of reality and pass from the realm of sound engineering into science fiction. The result is that life in space will not be what I describe here in detail. Why describe it then? Simply because I can thereby provide a sound base of departure. Where life in space differs from what is described here it will be better, cost less, and come sooner. New materials and techniques will make possible lighter, simpler, less expensive structures than those described here. There is the paradox — to get anywhere I must design things in detail, knowing well that they will not be like that — but it is a beneficent paradox. Things will either be like that or they will be better.

43

I have found it necessary to make one exception to the conservative policy of designing in only what we actually have. This is that power will be available from the combination of deuterium atoms to form helium. Without this we cannot get to the stars. There is simply not enough power in sight from any other source. The problem of power from deuterium is being intensively studied and worked on in the United States and Russia. The fusion of deuterium atoms and the heat produced by it have been observed many times but all equipment tested up to now requires more power to operate it than it produces. In common with others who have written on this subject I believe that the problems associated with building a productive fusion power plant will be solved within the next twenty years.

A typical structure which will be built and used in space is a sphere about two hundred and fifty feet in diameter and housing about a thousand people. The sphere will be in orbit about the earth. One face of it, consisting of a mosaic of glass windows, will be kept always turned toward the sun. This half surface is a greenhouse. The plants are in racks, their roots attached to plastic sponge to which nutrient solution is piped. There is constant motion as plants are brought forward for a few hours of bright sunlight and then withdrawn to rest in the shade. These plants consume the carbon dioxide exhaled by the people and restore the oxygen they breathe. These plants constitute the entire food supply of the station and also supply some chemicals and fillers used in plastics. The people live in multi-room apartments which have neither windows nor movable furniture.[2] There are recesses in which things can be stowed, a chute for dirty clothes, and another for trash. There is no cooking in the apartments; food comes hot and ready-cooked in packages something like TV dinners. To read or rest, one attaches one's self to the wall by a short length of cord. The people move about by swimming in the air exactly as we now swim under water. To move fast or go some distance they wear swim-fins like our present ones but larger and lighter. It is possible to get up to speeds of as much as twenty miles per hour with these but inadvisable because the fins

44

have little braking power. There is nothing corresponding to day and night and the people work and sleep in shifts from rather complicated schedules and calendars.

There will be several million of these stations in orbit about the earth, some smaller and some much larger than this one, none closer to the earth than five hundred miles, to avoid the drag of the earth's atmosphere, and none further out than one hundred thousand miles, to avoid excessive interference by the pull of the moon. Seen from a distance, this assembly would look much like the rings of the planet Saturn. You could make a telephone call from any station to any other, there would be interstation buses, and much travel between stations for business or pleasure. There will be television as well, and sports events in one station can be seen in others.

The air breathed in the stations will probably be oxygen plus a little carbon dioxide, at a total pressure of about three pounds per square inch.[3] The relative humidity will probably be kept near 70 percent and the temperature near 85° F. Clothes will be skimpy and will be worn for modesty rather than warmth. The choices of air temperature and pressure are moot and of first importance. The weight of the contained air of a station of this size is ten percent of the total weight; for larger stations the weight of the air is proportionately larger. The cost of building a station is roughly proportional to its weight. The Russians, for some reason which is not clear to me, use a mixture of oxygen and nitrogen similar to ordinary air. This would make their stations heavier and more expensive than ours. American practice so far is to use pure oxygen. There is a wealth of experience with this; pilots have been breathing pure oxygen at high altitudes for more than thirty years. So far no one has breathed pure oxygen continuously for more than three months at a time and there may be surprises when it is breathed for a year or more but there does not seem to be any fundamental reason why the mixture proposed here should be inadequate. One thinks of the danger of fire and remembers that three American astronauts were lost in such a fire, but that was on the ground and the capsule was filled with oxygen at fifteen pounds per square inch, not at

45

three pounds as proposed here. I have experimented at igniting cotton cloth in oxygen at fifteen pounds and at three pounds. At the higher pressure it burns explosively, but at the lower pressure it burns less rapidly than in ordinary air. Another factor to remember is that here on earth the action of gravity favors combustion. One sees air drawn to the base of a candle and stream out of the top. This happens because hot air is lighter than cold air. In space, in the absence of gravity, there would be no tendency for hot air to rise in cold. A candle lighted in space, even in pure oxygen at fifteen pounds per square inch, would probably go out. The carbon dioxide and water vapor produced by the combustion would choke it, and there would be no force tending to remove them from the vicinity of the wick. I think the danger from fire in space stations will be less than it is now in our structures here on earth.

Left to itself the greenhouse would get too hot for plants to grow or people to work there. The easiest way to cool it would be to let some liquid, possibly ammonia, evaporate there in closed pipes, and then to condense the ammonia vapors on the cold face of the station — the one away from the sun. This situation contains all the essential elements of a power station, which in fact it is. On their way from the greenhouse to the cold side the ammonia vapors would expand through a turbine driving an electric generator. The ensemble generates about eighteen kilowatts per man in the station, which is about what is used here in industry. In a space station electric power comes free — in fact it is forced on you.

Refrigeration comes free too. By regulating the flow of heat to specific patches of the metal skin on the dark side of the station, these can be made to take any temperature down to −440° F.

Occasionally the station will be hit by meteorites which will make holes in the outer skin, most of these less than one one-hundredth of an inch in diameter. The station air will leak so slowly from these holes that there will be plenty of time for people to come in and plug them without wearing space suits. Occasionally larger meteorites will make holes and compartments will have to be sealed off pending more elaborate repairs. The weight of air lost in these accidents will ordinarily be less than one millionth of the air in the station and will be made up by oxygen in granite brought up from earth. Present data concerning the distribution of meteorites in space by size and frequency are not very accurate but they suggest that a meteorite large enough to penetrate several compartments and cause casualties would hit a station about once in a hundred years, and a meteorite large enough to destroy the station once in a hundred thousand years. Still, the impact of meteorites may be enough of a nuisance to make people prefer to live in from the outer skin of the station. There will be windows in the outer hull where people may watch magnificent displays of stars, moon, earth, and other stations, but these probably will not be popular. People may prefer to see it on TV rather than take the chance of being stung by a meteorite.

At various points just inside the outer skin of the station there will be pairs of flywheels whose axes are at right angles to each other. These will be used to turn the station in such a way as to keep the greenhouse facing the sun. By increasing the speed of certain wheels toward the left you turn the station to the right. These wheels will turn sometimes slowly, sometimes rapidly, and will occasionally reverse the direction of rotation to accomplish their purpose. Students of physics will realize that there is a limit to what can be done in this way; after a long period of time it will be found that to accomplish the purpose some of the wheels would have to turn so fast they would fly apart. Long term drifts beyond the capacity of the wheels to correct will be dealt with by means of maneuverable metal sails attached to the outer hull of the station. Sunlight exerts a small but definite pressure, similar to that of wind on earth, and these space stations can sail to a limited extent, but the time to see any effect of the sails is measured in months or years.

Deep in the station is a layer of administrative offices much like those we have here but with some differences. Nothing can be left on desk tops — it would simply drift off — and paper weights would be of no use. It is doubtful that paper will be used at all; its function will be performed by audio and video tapes and television types of display. Files will be completely electronic; to get a record you simply dial it and see it on a tube in front of you. There will be some typing but it will be directly into electronic files and not on paper. Typists will be strapped in their chairs to get sufficient purchase to strike the keys.

The organization of a space station will probably be something like that of a present day ship, with a commanding officer and under him line departments including agriculture, manufacturing, education, and maintenance and repair, and there will be a staff for planning.

The main purpose and activity of the station will be to fabricate new stations, first for orbit about the earth and then around the moons and planets of the solar system; finally, to build interstellar ships. None of these structures could be built on earth; they are so fragile that they would be flattened by the force of gravity here. Each station will have a hangar with a large door through which sections of new stations can be pushed out. While these are being built the hangar will be kept full of oxygen at three pounds per square inch, the same as the rest of the station. When a section is completed the oxygen in the hangar will be pumped into tanks (it will be too valuable to be merely vented) and the door to the outside opened. The new station will be completed outside the old one by joining sections fabricated inside the hangar. This will be done by men in small airtight vehicles with viewing ports and external mechanical arms and claws capable of manipulating tools; not by men in space suits of the type we have seen used on the moon. These are too stiff and cumbersome.

Some of the manufacturing will be done in machine shops much like the ones we know, cutting and shaping metal and plastic parts, but with differences. There will be no cranes; anyone in the shop can move any part, no matter how massive; the problem will be to not get it moving too fast, in which case it might push right through a bulkhead. Lining up lathe and shaper beds will be less of a problem than it is here since there will be no gravity there to distort them. Metal and plastic chips from the cutting tools will be a problem because

48

they will not fall into the machine beds as they do here but will have to be caught and disposed of as formed. Lubrication of shafts and sliding surfaces will still be necessary, as it is here, because even in the absence of gravity, forces from cutting action will force these surfaces together. Spray of lubricants from shafts will have to be sucked away as fast as formed. Grinding operations will be especially troublesome because the dust will have no tendency to settle. Not that this will mean no dusting in shop and living quarters. Dust particles will become electrically charged, as they do here, and will be attracted to, and stick to, oppositely charged surfaces such as machine parts and walls.

Electricity will for the most part be generated and distributed at high frequency (because there will be no transmission lines and no stability problem) and at high voltage (because high vacuum, which is the best insulator there is, comes free in space). All but a few electric motors and generators will be electrostatic types which, though bulkier than their electromagnetic equivalents, weigh less than a tenth as much, watt for watt, and contain no copper, which will be a rare metal in space stations.

An appreciable fraction of the manufacturing effort will be devoted to tiny and extremely intricate electronic parts. Most of the workers in these sections will use microscopes and micromanipulators.

Considerable manufacturing effort will go to processing raw materials, brought from earth, by essentially chemical methods, avoiding the use of liquids as far as possible. Free liquid surfaces cannot be maintained in the absence of gravity. Where such surfaces are needed the operations will be carried out in centrifuges, operating in pairs turning in opposite directions to avoid turning the ship when the centrifuges are turned on or off. Air brought up from earth will be liquified by contact with surfaces on the dark side of the station, the nitrogen distilled off, and the oxygen used as station atmosphere. Part of the nitrogen will be combined with hydrogen to form ammonia, and this will combine with water and some carbon dioxide exhaled by the people to form plant nutrient

49

solution. The rest of the nitrogen will be vented. Twenty gallons of water per person will be needed in setting up a new space station, but to colonize the entire solar system will require water amounting to less than one part in a thousand of the volume of the sea.

Some of this sea water will be distilled and added to the station supply. Some of it will be electrolysed into oxygen, which will go into the station atmosphere, and some will be made to react with nitrogen as described above. Of the solids left after distillation the magnesium sulfate will be reduced to the metal for use in fabrication, and some of the sulfur will be used by the plants. Some of the sodium sulfate will go into plant nutrient solutions, together with all the potassium chloride recovered from the sea water. Most of the salt will go into the diet. Calcium sulfate will be used as a filler in plastics.

Granite will be fractionally distilled into silica, sodium, potassium, calcium, magnesium, aluminum, and iron oxides, plus a residue of heavy metals and their oxides. No container will be necessary; in the absence of gravity the melted granite will gather itself into a ball, and in space the high vacuum needed for distillation comes free. Most of the silica will be used in making glass for greenhouse windows and viewing ports; some of it, with all the sodium oxide and part of the calcium and aluminum oxides, will be used in making ceramic parts. Some of the calcium oxide will be reduced to metal which will be used in making parts never exposed to the station atmosphere. All the iron and the rest of the aluminum oxide will be reduced to metal for making structural parts.

Carbon will be needed as carbon dioxide to add to the station atmosphere and to be used in making chemicals and plastics. Some of this carbon will come to the station as limestone and some as coke. About five pounds of phosphorus will be needed for each additional individual established in space. Part of this goes into people's bones and part into the plant nutrient solutions. Part of the phosphorus will come up as the element and part as phosphate rock. The fluorine in the phosphate rock will go into tooth structure and to making plastics.

It will be impossible to choose the quantities of sea water, air, granite, limestone, and phosphate rock brought up from earth to the space station in such a way as to come out even. By the time one has enough of anything, say aluminum, he will have too much of something else. It seems probable that nitrogen, hydrogen, silica, and calcium oxide will come out in excess of requirements. The gases will simply be vented (from two jets on opposite sides of the station so as not to turn it). The solids will be put into plastic bags weighing a pound or less each, and these into larger plastic containers which will be projected downward into the earth's atmosphere. There the whole lot will burn up and the solids come down to earth in the form of a very fine dust. The cost of rejecting unusable material will be less than one percent of the cost of bringing it up from earth.

The plants in the greenhouse will consume all human wastes and the carbon dioxide breathed out by the people. Wastes will be sterilized and added to the plant nutrient solutions. The stench in the greenhouse will be appalling and agricultural workers there may have to wear equipment similar to that of scuba divers. The idea is most unattractive, but it is necessary. Human wastes have been used in China and in central Europe for a very long time. As I have mentioned in Chapter 2, I have lived in China for more than a year in places never free of this stench. I never got used to it but it could be borne. Plants break wastes down rapidly and completely, converting them into sugars, starches, proteins, fibrous materials, and oxygen.

The air in the greenhouse will be maintained at a pressure slightly less than in the rest of the station so that any leaks will be into it, thus confining the smell. Air going from the greenhouse into the rest of the station will be passed over hot copper oxide, which destroys all odors. The copper oxide is not consumed. Water vapor evaporated from plants in the greenhouse is carried with this air, deodorized, and condensed out for use in drinking and cooking. Air and water treated in this way will be completely odorless and purer than the air and water we get now.

Dead bodies will be cremated and the products purified

and returned to the station economy. Station trash and garbage will be handled in the same way. Thus all the material which comes into the station will be used over and over again, and if there were no increase in the number of people in the station and no new stations built, people could live there for hundreds of years with almost nothing brought from outside. "Almost" because there will inevitably be some leakage and some deliberate rejection of material from the station. Such wastage can be kept very small, probably less than one percent of the weight of the people per hundred years, if that turns out to be economically desirable. So long as raw materials are readily obtainable from earth, best economy will probably call for more wastage than this.

The reader may have wondered where, even in as large a station as this, room can be found for all the equipment and processes enumerated, and, even if sufficient space can be found, whether there would be enough people to operate all the different processes. The answer, of course, is no. In the interest of efficiency there will have to be considerable trade between stations, each of them specializing in one type of product, just as factories do here. The stations will be of different sizes, according to what they are making. There will be some huge ones, with hangar doors big enough to pass a whole station as big as the one described here. This commerce will be subject to some difficulties we don't have down here. Passing goods from one station to another one will ordinarily have to wait until they pass close by, which may take several months, and there will be a complicated scheduling problem. Stations located at Lagrangian points (see the note on these in Chapter 1) will not be subject to this difficulty. Stations placed close together there will remain close together, and it may be that each of these points will become the site of an industrial complex.

Housekeeping in a station will be minimal and consist mostly of swimming around with a hand net catching small articles which have come adrift. Food will all be prepared in a central kitchen because cooking in a condition of weightlessness is a difficult and complex process requiring equipment out of reach of a household. There will be, of course, no plates, knives, or forks, and everything, solid and liquid, will be squeezed into the mouth from plastic tubes. A variety of fruits will be available and can be peeled and eaten just as here. One will have to be careful not to squirt juice when eating a grapefruit, for example, because the juice would spread as a mist and get everything sticky. Most food will be mushes like present day baby food. Everyone will be a vegetarian. With the possible exception of stingless bees to fertilize plants it is doubtful that there will be any animals in a space station because of the difficulty of housebreaking animals in a weightless condition. Besides, as noted in Chapter 2, production of meat by animals is a wasteful process; a cow eats about ten times as much grass as she produces meat. It seems probable that second generation space people will read about our habit of eating meat — or, on TV, watch us eat it — with loathing. It may seem to them like cannibalism. When, perhaps five generations from now, the first immigrants touch down on a planet of a distant star they will, to their disgust, have to go back to eating steaks. Sheep, goats, and cows from earth will be able to live on the plants of a new planet, but the immigrants won't. All in all, eating in a space station will probably not be as enjoyable as it is here, and may become less of a social function. Alcoholic beverages will be easy to make in space and will probably be widely drunk and enjoyed there.

Social life in a space station will be much like it is here. People will sleep, work, eat, and watch TV *en famille* just as we do, with occasional parties and sports events. There will be libraries with science, literature and music available, but no one will go to them; you will dial what you want and if it is a book it will appear, page by page as you call for it, on a TV screen. If you want the book read aloud, or want music, it will come through earphones as you dial it. There will be doctors on call for the ill and hospitals much like our present ones.

Sports will differ in detail from those we have now because of the absence of gravity, but the ideas and competitive spirit will be the same. One such will probably be a three-dimensional football played something like water polo. Another might be badminton in which you have to get the bird through a hole in the net. Sports areas will be completely enclosed by nets capable of stopping a player gently if he gets going too fast. Spectators will be above and below, at both sides, and at either end of the play space. Weird, perhaps, but quite as enjoyable as our present spectacles, and the game will never be rained out. There will be interstation competition and sports stars. Theatricals in space will offer new techniques and very likely will be popular. There will be news commentators to tell us just how it all is, and preachers to explain why we must not do what we want to do. Superficially life in space will be very different; practically it will be much the same as now.

What I have been talking about is the end product of the evolution of space stations and the culture within them. All the people will have been born there, or will have come up from earth at an age of less than six weeks. They will have known no other life, and as long as the surface of the earth remains habitable they will watch with amazement, on TV, the strange ways of the people here, and the even stranger animals.

The problem of how to build these large space stations from a rocket base on earth is inextricably tied to the effects of weight and weightlessness on people.

All that is really known now about these matters is that some people can stand weightlessness for as long as three months, and that in these it has produced physical changes including partial decalcification of the bones. Physical changes are to be expected from so drastic a change in a way of life. The problem is how to use rather than combat them.

It is certainly not to be expected that every healthy person can adapt to weightlessness. Apparently at least one Russian has failed to adapt and died in orbit. There is nothing astonishing about this. Some people are so sensitive to motion that they cannot even ride in an automobile. Many others are unable to adapt to the motion of a ship at sea. These things are true but unimportant; what matters is that most people

54

can ride comfortably in automobiles and that probably more than half the population can adjust to the motion of a ship at sea. These are the people we are interested in. It seems a fair assumption that a significant fraction of the population can adapt to weightlessness for a lifetime. Ten percent will be ' enough to ensure the success of the program.

So far no American astronaut has had difficulty in adjusting to gravity on his return to earth but this will probably become more difficult the longer one stays in space, and it is likely that after a certain period of weightlessness, different for different individuals, this adjustment will become impossible. This seems to be a matter of indifference. If a man wants and is physically able to come back, he comes back. If he wants to stay out there, or if it is physically impossible to make the adjustment, he stays. It is likely that some people, after doing well in space for a while, will be suddenly hit by it, and will have to be brought back to earth to save their lives. It seems equally likely that some people about to die on earth from overweight or heart difficulties can be saved by sending them into space.

Only relatively small space stations, not much bigger than Skylab, can be built on earth and blasted into orbit. Larger structures would collapse of their own weight if we tried to build them here. We must begin by building a number of small, cramped, uncomfortable stations and sending them up as barracks for construction workers. These will be somewhat analogous to the covered wagons of one hundred fifty years ago. They will not be self sufficient. All the food, air, and other supplies must be sent up to them. These construction workers will be rotated back to earth as their physical needs and contracts require. Presumably the periods of duty in space will gradually be increased, and readjustment to life on earth more difficult. The day will come when some oddball will decide not to come back but to live the rest of his life out there.

As tours of duty in space become longer the physical changes due to weightlessness will increase and require attention, some of it medical and some of it in the form of special exercises, but the goal will not be to fit people for return to earth but to fit them to stay in space. There will be no need there for great physical strength. To get things done one will have to push a little longer, not a little harder. If decalcification can lead to flexible arms and legs like those of a baby, fine. (Deliberate decalcification of bones to eliminate radium from the body leads to brittleness.) If long exposure to weightlessness leads to atrophy of certain muscles, say of the arms or legs, we need to get our man up there while he is still young enough for his skin to shrink without wrinkling, but otherwise we are unconcerned. If his chest expands a little to deal with the thinner air, and if this, combined with shrinkage of the arm and leg muscles, makes him look like Popeye the Sailor, this is fine too.

After we have a few dozen people up there, and a shuttle service working, and emergency stores of food and oxygen in place, we shall build complete parts of space stations on earth, each small and strong enough to be lifted into orbit by present day rockets. The people up there will bolt these parts together and move into new, more comfortable quarters. Then, from sections brought up from earth, they will build a hangar large enough to house and pass through its doors space station parts too big to send up by rocket. From these still larger space stations will be built. Some of these will have greenhouses and be partially self sufficient. Experience with these will lead to larger and better designs until the fully self-sufficient space stations described above are achieved.

Selection and training of personnel will make or break the space program in the years immediately ahead. We need tests to find out at the earliest possible age whether the candidate can endure and prosper under weightlessness. If he can he then needs to be trained in the precise but not necessarily elegant use of language. His life and that of those with him may depend on his instant and exact understanding of an order given to him. Then he needs to learn basic science with emphasis on physiology. After that he needs endless drill on survival techniques and emergency procedures. Finally he must be taught some technical or administrative skill which may be quite narrowly focused — electric welding in high vacuum or the use of adhesives, for example. Then he is ready to go. I realize that for a time we shall have to send middle aged men into space, as we and the Russians do now. The reason is that for the next few years we shall have to send out men highly competent in a range of subjects too wide to be mastered under the age of thirty. A younger man just hasn't lived long enough to acquire this competence. But a construction worker needs no detailed knowledge or experience of piloting, astronomy, mechanical engineering, the problem of motion in a gravitational field, the operation of computers, and so on. All he needs to know is how to fit two parts together right, plus some emergency procedures. This knowledge can be acquired at an early age and it had better be, because the early space construction workers probably won't live very long. The hazards to which they will be exposed will be just too great.

There is an historical parallel. In the early eighteen hundreds, life at sea in naval vessels was so difficult and dangerous that officers had to go out as midshipmen at the age of twelve in order to acquire enough experience to be useful before they passed through the peak of physical endurance at about age twenty five. Few of these officers lived to be forty but they performed splendidly while they lasted. Living in space during the pioneer days may turn out to be a more extreme example of this situation.

It can be confidently assumed that children will adapt more easily than adults to weightlessness, and if need be to return to the gravity of earth. This suggests that recruitment should be done at the age of six or less, and that most of the training be done in space. In that way, those who are found to be unable to endure weightlessness can be returned to earth while they are still young enough to adapt to weight easily and to learn the skills necessary to life on earth before they become adults. In fact, I believe that some of these recruits, orphans or unwanted babies, should be sent into space before they are six weeks old. That way they will, a few years later, have no memory of weight and be able to adapt to space life that much easier.

When the first baby is born in space the program will

enter a new phase of decreasing dependence on earth. Conventional genetic theory tells us that the baby will be exactly the same as one born here but this is clearly not true. Identity at the instant of conception, yes, but not beyond that. In space the mother's chemistry and physiology will differ significantly from what they would have been on earth, and under weightlessness the circulation and probably other physiological conditions of the fetus may differ substantially from what we have here. By the time the baby is born it will have already gone a long way in the process of adaptation.

Some miscarriages which would not have happened on earth may be expected in space, and some miscarriages which happen on earth would not have happened in space. Some children born in space will be unable to adapt to weightlessness and will have to be sent to earth to save their lives. Even so, one cannot doubt that people born in space will, on the whole, do better there than people born on earth could possibly do, and that second and third generation space people will be able to do things quite impossible to anyone born on earth. They will be able to last out and endure journeys to the stars.

Space stations for orbit about the moon and placement at the Lagrangian points of the earth-moon and earth-sun systems will be identical with those described above. For a while, at least, these will be constructed in orbit about the earth and towed to location. Stations for orbit about Venus and Mercury will have sunshades to keep them cool. Instead of full windows the greenhouses will have only ports, and there will be optical diffusers inside so that the plants will not be exposed to the full glare of the sun. A tail in the shadow will radiate the excessive heat which the head will pick up. Stations for orbit about Mars and the outer planets and their moons will center on the focus of parabolic mirrors and the greenhouse windows will cover the entire surface except for the hangar doors.

The moon[4] will be provided with a set of mirrors in orbit (for details of these mirrors please see appendix 6) which will deflect the heat of the sun away from it or toward it in such a way as to keep the surface temperature about the same as that of earth. Nuclear power will be used to decompose the lunar rock and provide an oxygen atmosphere, which will shield the people on the lunar surface from being stung by meteorites and from the ultraviolet light from the sun. Whether enough water can be had from the lunar rocks or whether it must be brought up from earth is moot, but water will probably be in very short supply on the moon. It may be possible to have outdoor swimming pools but no rivers or lakes. Permanent residents will go directly to the moon from earth and will not have to adjust to weightlessness. When these changes are completed there will no longer be a crescent or a full moon; it will show a complete disc of pearly grey color throughout the month.

Piecemeal developments on the moon will be uneconomical; it must be done completely or not at all. The cost of real estate there will be about half a million dollars an acre, not too different from what you have to pay for some real estate on earth.

Mars will be provided with mirrors in orbit to deflect more solar heat to it, first to melt the polar caps, which will provide some rivers and lakes, and then to heat it up to about the same temperature as here. Then plants we bring will grow there and convert almost all of Mars' carbon dioxide to oxygen. This will take about a hundred years. This may have to be supplemented by oxygen[5] from the planet's rocks to produce an easily breathable and protective atmosphere. Then people can live there under conditions much like those here on earth. Similar procedures will also make some of the moons of the outer planets habitable.

According to an article by William N. Agosto in *Harper's Weekly* for February 21, 1975, Professor Gerald K. O'Neill of Princeton has been thinking and lecturing along similar lines. His plan for the colonization of space differs from mine in that his first space stations would be in orbit about the sun rather than the earth, he would get his minerals from the moon and the asteroids rather than from the earth, and he provides artificial gravity by rotating his stations while I assume operation under conditions of weightlessness. His economics and timing appear to agree with mine.

Notes

1. There is nothing exotic about the ICBMs we are presently deploying.

2. No furniture because, one, it wouldn't stay put unless bolted down, and, if bolted down, would be a source of frustration because you couldn't rearrange it; two, you don't sit or recline in space — you float, even when sleeping, and so don't need furniture; you don't write with pens or pencils, which would be forever floating off, but type everything on a keyboard set in the wall, with belts to hold you near it — otherwise you would be pushed backward or upward when you struck a key; and four, since you do no cooking and use no dishes — everything you eat or drink you squeeze into the mouth from collapsible, disposable plastic tubes — you don't need tables, stoves, or refrigerators. There will be mirrors set in the walls, and pictures on it (screwed down, not hung), and recesses in the walls, probably with sliding panels, for books and clothes. Bookshelves would be of no use; the books would simply float off them.

3. At sea level, the air we breathe exerts a pressure of 14.9 pounds per square inch and contains twenty percent by volume of oxygen, seventy-nine percent of nitrogen, and the balance is made up of argon, water vapor, and carbon dioxide. It would be extravagant to breathe ordinary air in a space station because it weighs five times as much a cubic foot. So far as is known, the nitrogen in the air plays no part in the body chemistry, so we leave it out to save weight. The oxygen taken in per breath from the atmosphere proposed here would be the same as from ordinary air. We would like to use still less oxygen, and have a still lighter station, and we could get by with one and a half pounds per square inch, corresponding to what we breathe here at an altitude of eighteen thousand feet. (I have ridden quite comfortably over the Himalayas

without oxygen at an altitude of twenty-three thousand feet), but we have to leave a factor of safety so that if there is an accident to the station and the pressure begins to fall there will be time to do something about it before people start passing out.

The design temperature of 85° Fahrenheit was chosen to conserve body heat. The less heat you lose to the environment the less food you need. This is about the temperature deep in the tropics in early morning and late afternoon, and I know from my own experience that, once you get used to it, it is the most comfortable temperature there is. The design humidity — 70 percent — is a compromise; make it a little more and things get soggy, make it a little less and your nose gets dry and you may catch a cold.

4. We may wish to save the moon for a special purpose and delay colonizing it. It will be impractical to place large astronomic telescopes in space stations because these are so light, and so subject to wobble as people move around in them that it will be impossible to hold the telescope on target with the required accuracy. It will not do much good to build larger telescopes here on earth because the atmosphere shimmers and blurs the pictures, and lately because of the glare of cities on the night sky. The back side of the moon seems an ideal place to put a very large telescope. There are no trucks to make the ground tremble there, as yet no atmosphere to obstruct the view, and the body of the moon will be interposed between the glare of the earth and the sky. Such a telescope could do wonders during the next hundred years — it might even make out the planets of some of the nearer stars and so improve the chances of colonization from interstellar ships — but in a few hundred years it would be superseded by much larger telescopes in orbit about the sun but very far out. Then we can go ahead and develop the moon.

5. Some of this oxygen will escape the gravitational attraction of the moon and Mars and drift off into space. As nearly as I can calculate it, this loss will amount to about one percent of the entire atmosphere in ten thousand years. This loss can be made up by decomposing more rock. The earth's

atmosphere has been losing oxygen in this way for about three billion years but there is still plenty left.

4. Interstellar Ships

Dr. Edward Purcell, professor of physics at Harvard and Nobel Prize winner, lecturing to the Staff of the Brookhaven National Laboratory about interstellar travel, said "Well, this is preposterous, you are saying. That is exactly my point. It *is*-preposterous." Dr. Glenn Seaborg, also a Nobel Prize winner and formerly Chief of the United States Atomic Energy Commission, implies in various speeches and writings that he believes that interstellar ships can be built and flown. Dr. Purcell was thinking of round trips to the stars at speeds more than half that of light. My own calculations and those of my students agree that the thought of such voyages is indeed preposterous. They would burn too much fuel, and it would be impossible to shield the crew adequately from the nuclear radiation produced by the ship's engines and by impact with stray atoms and meteorites in space. Dr. Seaborg was thinking of one-way trips at much lower speeds. There is enough deuterium in the sea to fuel more than a hundred thousand of these, and the shielding problem, though it has to be carefully thought out, is solvable using present-day materials and techniques. The cost comes out to about two billion dollars per ship (see Appendix 2).

An inter solar system ship powered by nuclear engines and taking off from orbit about the earth will arrive in orbit about Mars with about 70 percent of its takeoff weight. An interstellar ship taking off from orbit around the sun will arrive in orbit around a star weighing less than one one-hundredth of what it did. The rest of its takeoff weight is made up of fuel, fuel tanks, and engines.

63

The chemically fueled rockets we have watched on TV taking off for the moon are adequate for interplanetary flight within the solar system, but only just — they are quite incapable of interstellar flight. The nuclear engines which will drive interstellar ships do not have enough thrust to take off from earth and must be launched from orbit. It is unlikely that a nuclear engine capable of lifting itself off the earth can be built. To get materials and men from earth to orbit we shall have to use chemical rockets for a long time. For interplanetary flights within the solar system the nuclear engine is superior in performance and for interstellar flight, essential.

The takeoff of a chemically-fueled rocket is spectacular. The flame is dazzling, the noise incredible, and the push and vibration close to what the hull and passengers can stand. You would hardly notice the departure of a nuclear-powered craft from orbit. The flame will be barely visible, there will be no noise, and the push and vibration will be less than in an automobile pulling away from the curb. In a chemically powered rocket the push begins at about twice that of gravity and builds up to about five times that as the fuel is burned. To get a nuclear ship up to a tenth the speed of light in two years requires a push only 5 percent that of gravity.

The complex of rockets, fuel tanks, and vehicles making up a caravan to a star will be assembled in a straight line about fifty miles long. The harness connecting the units will be made up of triads of thin-walled metal tubes with triangular spacers at intervals along the traces. These tubes will carry fuel and electric power from one unit to another and will also serve as wave guides connecting the computer of the system. The caravan will be assembled at a remote point of the solar system, far from the sun and any planet, where tidal forces will be much less than they are in earth orbit, and where the heat of the sun is so feeble that liquid hydrogen, liquid deuterium, and liquid helium-3 can be stored without excessive refrigeration.

At the head of the caravan is the largest of the rockets, whose multiple jets are angled out slightly so as not to blow on the following tanks. This rocket is followed by its fuel tanks, these by a second rocket and its tanks and so on, until one comes to a bumper,[1] followed by living quarters for the coast phase, followed by a bumper, followed by the living quarters used during the slow-down phase, followed by a bumper, followed by the tanks of the last rocket, and ended by this rocket itself, which is headed to fire in the opposite direction from the others. When the first rocket is spent it is jettisoned with its tanks, the second rocket fired and then jettisoned, and so forth, until the next to the last rocket. When this is spent the remainder of the caravan is up to a tenth the speed of light. The next to the last rocket is not jettisoned after firing but its hulk is left out in front during the coast phase to act as a bumper. At the end of the coast phase the next to the last rocket and the crew quarters of the coast phase are jettisoned and the last rocket fired, bringing what is left of the caravan down to nearly zero speed in orbit about the target star.

During the coast period, which may last several hundred years, the crew and their descendents will live in comparatively spacious quarters designed for living under conditions of zero gravity. These will contain neither greenhouse (the sun is now too far away to furnish enough light and the destination star not yet close enough to do so) nor garden (one in the less spacious quarters designed for living in a gravity 5 percent that of earth, during the slow-down phase, will supply the food and oxygen needed during the coast phase). At the end of the coast phase these quarters, together with the hulk of the next to the last rocket will be jettisoned. It would cost too much to slow them down.

During deceleration, which will last two years, the crew will live in the smaller quarters with the garden which, operating under artificial light, will supply food and oxygen. After the landing party has gone these quarters will slowly be converted, working entirely from within, into an orbiting station with a greenhouse growing food by the light of the host star.

There will be no windows in interstellar ships because there will be nothing out there to see. The fore-and-aft axis is obstructed by rockets and tanks — one cannot look ahead to see the target star or back to see the sun. This introduces a complication because course corrections must be made about a month before the end of the acceleration phase and about a month after the beginning of deceleration. The coast phase quarters and those used during the deceleration phase will each be equipped with a telescope looking out through a side port at a mirror far enough out to look past the rockets, supported there by a mast. By rotating this mirror one can see either the target star or the sun and the angle between the lines drawn to each of them accurately measured. The telescope-mirror system will also be used to search for planets of the target star. Not until the colonists are within three months of destination can they expect to see these. Then will come anxious moments. All machinery will have to be shut off and everybody hold quite still, scarcely breathing, to keep the ship steady enough for reliable observation. If the star has planets the plane of their orbits and their direction of rotation around the star must be determined in time to make course corrections so as to come in in the orbital plane and in the direction in which the planets are revolving. The ship will then be put into orbit, first around the star itself and then around the selected planet.

Communication with earth during the voyage will not be simple. Conversations will be impractical because even at the instant of launch it will take more than an hour for the voice to make the round trip to earth and back. By the time the ship reaches its destination the time from question to reply will be more than nine years. Transmission will be by a dot-dash method or something similar and will be very slow because as you decrease the speed of transmission you also decrease the power required to get an ungarbled message through. In the solar system, out well beyond the planets, to decrease tidal forces and the temperature of the metal, there will be a radar dish type of antenna several hundred miles in diameter, made up of a spidery web of girders and a mesh of wire less than a thousandth of an inch in diameter, which will receive messages from the ship and transmit messages from earth. The ship will have a small radar-type dish antenna far enough out on a mast that it can transmit and receive past the rockets. It is doubtful whether any communication will be possible during the acceleration and deceleration phases because of radio interference generated by the rocket engines.

Getting an interstellar ship from the earth to a star is a social rather than a technical problem. Since the voyage will take at least fifty years the landing will be made by a third or later generation born on board. There will be no police in an interstellar ship, none waiting on arrival, and no practical penalty which can be imposed on misconduct. Crew members can be restrained from fighting and killing each other off only by a religious sense of mission and by exceedingly wise and forceful leaders. Probably most of the ships launched will be lost because of social conflicts, but this will not matter because enough ships will be launched so that a few will get through. Still, it will be worthwhile to minimize social casualties.

This will be done by careful selection and training of the crew. In spite of the cliché "blood tells," inherited ability is the most important attribute a man can have. In choosing crew, genealogy will be scanned. But not many people — probably less than one in a hundred — of those who have the right heredity can be taken. Selection will have to be made on the basis of tests and there will be not much time for these for the captain should be under thirty and the crew under twenty when the ship sails.

The captain need not have specialized technical knowledge (the crew members have that) but he must be acquainted with the great thought of the past and be the only one on board who is. So long as the crew understands that he knows things which are forever beyond them they will respect and trust him. By the time the first generation born on board is ten years old the captain must have chosen his successor and begun to train him. The last flight captain will have to train two, one to lead the landing party and one to command those left behind in orbit about the host star. If the captain dies before his successor is trained the ship may be lost.

The essence of this kind of leadership is to spot and do something about personal discords before any of the parties to them know that they exist. Even given this genius for leadership, and all the experience in the world to draw from, the task will, in most cases, be impossible. In this situation where everyone knows that he will never again see anyone but the associates he now has, and that there is no possible way of escape, the most trifling thwarted wishes and slights will become of decisive importance. Perhaps merely a choice of cabins. Perhaps irritation with the timbre of voice of one associate. Perhaps a feeling that his job is being downgraded in importance. Perhaps a tendency toward wife-swapping or homosexuality. Any such feeling of discontent could grow to the point where it endangered the ship. Logically the captain hasn't a chance of success — he can get through only by luck and statistics.

A great deal can and will be done to minimize social difficulties by selection of the members of the crew for compatibility, first as couples and then as close neighbors. The larger the crew the less the social difficulties will be, but the larger the spaceship will have to be too, and the fewer the ships which can be sent out. A social-economic balance will have to be struck. The lower limit of crew size will probably be set by genetic considerations, though egg and sperm banks will be carried in case of need. My present guess is that the initial crew should consist of four couples.

Note

1. These bumpers are circular discs up to six feet thick. Their purpose is to shield the crew from the mechanical effects and nuclear radiations from the nuclear engines, and from impact with the stray atoms, dust particles and grains of sand in space. The material of the bumpers is chosen to minimize the production of radiation by impact of these materials. For the theory underlying the design of these bumpers see Appendix 2.

5. Landing

The size and composition of the landing party is critical. If there are too many, supplies must be sacrificed to accommodate them, and if there are too few the party may become helpless if two or three become sick and die. Four couples may again be the best compromise — one pilot who is also a mechanical engineer and electronics man, a chemist, two doctors, and four farmers. Of all the different conditions which may exist on colonizable planets one, which may be fairly typical, will be considered here. Let us suppose that this planet is somewhat smaller than earth, has a surface gravity of two thirds as much, has a somewhat shorter day, and no moon. The average temperature is a little less than that of earth and there are large polar ice caps. Most of the surface is ocean and the only inhabitable land is islands near the equator. The atmospheric pressure is half that of earth and the air is mostly oxygen, with a little nitrogen and carbon dioxide. The most advanced plants are mosses rising no more than an inch above the ground, and the only animals are clams, snails, worms, and insects. And suppose finally that the landing was made about two hours after sunrise on a clear day on level ground about twenty miles from the ocean and a few hundred feet from a stream.

The landing craft came down with quite a bang, but no bones were broken, although everyone was badly shaken up. The landing craft immediately began to heat up in the sunlight. A hatch was opened and the colonists took their first breaths of the new air. The smell was penetrating and disagreeable. Almost immediately some colonists were bitten by insects, and the allergic reaction to these bites was violent. The effect of gravity, though anticipated and experienced in mild form two years earlier during deceleration, was debilitating and almost insupportable. Motions were awkward and no one was able to stand up. The immigrants had to crawl about as they tackled the most urgent task which was to get a mosquito net up and insect repellents on. By the time this was done, about nine in the morning, they were beginning to sunburn. The next thing was to get a tent up. A simple thing, which any American teenager can do easily, but dreadfully hard for these people unaccustomed to working under gravity. By the time the tent was up, about ten, they were thirsty and one of them, wearing anti-insect clothing, crawled to the edge of the stream and brought back water. It tasted right, and was refreshing, but they had trouble learning how to drink out of a cup after so many years of sucking water out of plastic tubes. These things accomplished, they napped through the heat of the day, becoming adjusted to the strange smell of the air. Toward evening the farmers crawled to the edge of the stream, cleared away the moss on a small strip of damp earth, cultivated the ground, and planted different kinds of seed of fast-growing edible earth plants, including radishes, peas, beets, and potatoes (these will grow from seeds though ordinarily sections of potato are planted). Back at the tent the chemist had opened some dehydrated rations and was soaking them in water from the stream. The pilot had transmitted an account of the day's adventures to the ship orbiting overhead, which immediately began transmitting it to earth, some ten years away.

By morning three members of the party were quite sick, apparently from allergic reactions, and only one — fortunately a farmer — was completely well. He tended the others, making them as comfortable as he could, under instructions from the doctors, and prepared more of the dehydrated rations. All but the two sickest were able to eat something and keep it down. One of the doctors felt well enough to help out. The farmer who came through best attempted to stand but couldn't quite make it.

70

By the end of the third day all but one of the farmers were able to get around. The chemist began to analyse the moss and found no alkaloids. He then started to sort out the moss proteins, testing them out on patches on his skin. No alarming allergies appeared and one of the farmers began to eat small amounts of the moss under the observation of the doctors. It was nauseating but he managed to keep some down. The chemist went to work on the worms but had such strong allergic reactions to their proteins that he abandoned them and tried some insect grubs which proved edible. He found many bacteria, apparently none of them harmful.

The fifth night there was a violent storm and the tent blew down. Without lights and unable to stand they could do nothing about it and were soaked and chilled. By noon of the following day they had the tent back up but the exposure was too much for a sick farmer, who died. He was buried, with great difficulty, by evening. Most of the seeds planted the first day were washed out but some beets survived and sprouted, carefully tended by the surviving farmers. Other kinds of seeds were planted and more than a hundred square feet set to beets. They began to collect worms, chopping them up and using them as fertilizer. All human wastes were applied to the beet field. The chemist anxiously doled out the rations brought from above, stretching them with moss and insect grubs to the extent the immigrants' stomachs could take it. Everyone was losing weight but the farmer who came though best could walk a few steps now, after which he was utterly tired out. None of them noticed the bad smell of the air anymore.

By the end of the month all the survivors could walk and several had stopped losing weight. Some radishes had been picked and eaten. A few stalks of corn were growing and potato plants flourishing. The insects did not attack these strange plants.

By the end of the second month they were eating mostly beets, with some radishes and peas, and some moss and grubs. Protein was still coming mostly from the stock brought down with them. The toughest farmer had begun to gain weight and could walk half a mile. The chemist had begun to extract and crystallize sugar from the beets. The corn was tall and ears had appeared. Plans were being made to move the camp to the mouth of the stream where it flowed into the sea but no one had been there yet. It was decided not to move until the corn had been harvested and dried.

Everyone took turns cranking the hand generator which charged the batteries for the radio and there was regular communication with the ship orbiting above, but this was of little help except for weather forecasts based on what the ship could see of the cloud cover. The dwindling supply of protein brought down with them was anxiously watched and rationed. It was doubted that the protein from insect grubs and the little to be had from their crops would keep them healthy when their stores were gone. The hope was protein from the sea.

Two farmers and one of the doctors started off toward the sea carrying light packs. After going two miles they had barely enough strength left to put up a sun shade. Insects, though still a problem, were not as bothersome as they had been, for the immigrants had developed a partial immunity to their proteins. Even so, the advance party was badly bitten before it had rested enough to put up a tent and screens. They ate dried corn and insect grubs plus a little protein from the ship and some sugar from the beets, but it was two days before they were rested enough to make the trip back to the base camp, leaving a small hoard of stores at the advance camp, and they had to rest two days more before starting back toward the sea again. This time the third farmer and the chemist went along with them carrying packs of supplies. The amount which could be carried on each trip was so small that it was six months before the entire party arrived at the sea.

The beach was dazzlingly white and the water clear, though cold. The rise and fall of the tide was only a few inches. There were a few plants, some attached, some floating, and it was obvious that the factor limiting their growth was lack of nutrients in the water. The beach was made of ground and broken shells and these were found to contain a very small amount of phosphorus. There were a few shellfish, some poisonous, some not, and the edible ones proved a welcome change from insect grubs, but the effort required to collect them almost balanced out the strength they gave.

The colony very nearly didn't make it. The low point came when the protein brought with them was exhausted. There was a farm going at the mouth of the stream but the group had barely enough strength left to work it. All had again lost weight. One more had died. But then, gradually, strength returned. They were getting accustomed to working under gravity. Muscles previously latent began to develop where they would do some good. Coordination improved. They were able to eat more moss. A baby was on the way. It was time to think about animals to be sent down from above to improve their situation.

Yaks seemed the best choice. They would probably be able to live on moss alone, extracting the very small amount of protein it contained and concentrating it as meat. Their wool, which they would not need in this climate, could be woven into cloth. The people on the ship aloft concurred. Yak eggs and sperm were taken out of the bank and ten yak embryos started in plastic jars. Six of these came to term. Timing was critical. The little yaks were fed a milk substitute for a week and then sent down in a small landing craft, together with a month's supply of dried milk substitute.

It was an anxious time below. The craft landed less than ten miles from the camp and the strongest farmer managed to reach it in time to save five of the yaks, two males and three females. One of the females died a few days later. The surviving animals were soon nibbling at moss and corn shucks. One of the males was slaughtered when it was six months old and the colonists had their first taste of meat. They did not like it but the good effect on their health was evident.

The first baby was born, a boy. He was not weaned until he was a year and a half old, and then went on a diet of yak milk for he had difficulty adjusting to moss and shellfish, but finally he managed. He learned to walk at nine months and was soon frolicking around the camp in a way which amazed the tired, weak colonists. By the time he was six he could out-hike any of them and when he was ten they relied on him to find things sent down from the ship. By then more babies had come along and all of them lived, helped by a diet of yak milk. By this time the herd had increased to over fifty, and one was slaughtered every month. Yak butter and cheese became staples of the colonial diet and cloth was being woven from yak hair.

It had become clear that the most useful thing to do next would be to acquire pack animals. Donkey foals were sent down and raised with some difficulty. Geese were an easier proposition. Eggs were laid in the mother ship and sent down to be hatched, and there was cracked corn waiting for the goslings, who could eat moss from the start but needed a grain supplement until they were grown. Since there were no seasons at the location of the camp the geese were likely to lay at any time and eggs found their way into the diet as novelties.

It was desired to harvest the insect population and this called for a difficult choice. Insect-eating birds fly well and would simply move out of reach. They decided to try lizards. In the absence of natural enemies these throve and became so numerous that a fence had to be built to keep them out of camp. The colonists learned to like lizard meat. Dragon flies were also brought in and the number of stinging insects was greatly reduced. Within a few years a balanced economy set in near the camp, with the lizard population limited by the scarcity of insects and the insect population limited by the surviving lizards. Far from the camp there was no such equilibrium and the lizards advanced in great numbers into new territory, eating up the insects as they went.

Twenty years after the landing the population was about two dozen and life had become much easier. No one ate moss or grubs any more and attention turned to better dwellings. There were no trees on the planet and the rock outcrops were difficult to work. Bamboo, coconut palm, and papaya seeds were sent down and grew well in the lime-like soil. Fast-growing mahogany and mango trees were also cultivated, and by judicious use of human and animal fertilizer and the ashes of burned moss, were enabled to survive and propagate. Wooden houses began to appear.

A little more than fifty years after the landing, the last of the original colonists died. The population had grown to be over a hundred. It had become a pastoral meat-eating society. Yak herding was the principal occupation and some of the animals grazed more than a hundred miles from the town. Although the planet had not yet been explored it was doubted that deposits of phosphate rock would be found. Shortage of phosphorus might limit the population which could be achieved and the number of interstellar ships which could be launched. The soil contained only a few parts per million of phosphorus. In moss this was concentrated more than a hundred fold. Yaks, eating the moss, concentrated the phosphorus still more in their bones. It would require phosphorus from the bones of one yak to establish a man in orbit. The colonists began stockpiles of yak bones. The yaks also concentrated the small amount of nitrogen present as proteins in the moss. Most of this they dropped as dung which was dried and brought on donkey-back to fertilize the fields. Lizards, chasing after the diminishing insect population, were caught, dried, chopped up and spread on the fields for the same purpose, as were the ashes of sea plants. Corn and potatoes were now the staple crop, augmented by coconuts and fruits. Carp had been loosed in the mouth of the river and did well there. They were occasionally eaten but more often used as fertilizer on the fields.

The colonists were in constant communication with earth via the ship orbiting overhead, and were receiving indispensable technical information despite the ten year time lag. Two other colonies on planets of distant stars were profiting from their experience.

By this time less than half the labor force was needed to produce food. Ships were built, all sailing vessels. The entire planet, right up to the polar ice caps, was explored for minerals. No phosphate rock was found. There was a little silver, a little gold, some lead, tin, and zinc, some manganese and

75

chrome, but only low grade iron ore. There was no coal and no petroleum.

The sea was only half as salty as on earth, somewhat richer in potassium but poorer in phosphorus. It contained a higher concentration of deuterium than the waters at home.

The chemists produced alcohol by fermenting sugar from beets and oxidized this to acetic acid. They got citric acid from fruit juices. In the sea there were some iron-concentrating algae. The iron was dissolved out of these by the acids and precipitated as a pure ore by the addition of an extract of wood ashes. Clay deposits had been found and used to produce pottery; this clay was now used to make crucibles in which the iron ore was smelted with charcoal from coconut shells. Carbon was washed out of the crude iron under a flux of lime and crushed rock; the resulting steel could be forged and hardened. The colonists could then make their own cutting tools, replacing the few brought down from above and now worn down to nubs. Soon they were building first crude and then sophisticated machine tools, run by donkey power.

Two hundred years after the landing the population exceeded two hundred thousand, of whom less than a fourth were in agriculture. There were steel sheet and rolling mills powered by steam engines, using palm logs as fuel. Vehicles were still mostly donkey-drawn but there were a few battery-powered cars. The first electric lights were burning. Plastic parts were being molded.

Three hundred years after the landing the population exceeded ten million. Solar energy was being used for power, some of it converted into electricity by light cells, and some of it used to superheat water which was stored under ground for running electric generators at night and on cloudy days. Deuterium had been concentrated from sea water and the first thermonuclear power plant was about to commence operations. Magnesium and aluminum were being produced in large quantities by electric decomposition of rock. The last donkey was gone and there were many electric cars. The whole usable land area was now being grazed and the yaks were getting more seaweed than moss. Computers were being made and used. The first rocket capable of putting a satellite in orbit was almost ready to go.

Three hundred and fifty years after the landing the population was more than a hundred million, part of it living in orbit above the planet. The thirty-odd people still living in the interstellar ship had been rescued and the ship itself scavenged for metal. The genetic bank had been used to start a zoo at the principal city of the planet. Radio communication with earth was now by means of a huge dish in orbit way out from the planet; the increased size of the new dish and the greater power available speeded up the rate of transmission and reception. This was fortunate because the planet now required an amount of technical information from earth which tended to saturate the facilities. On the planet the moss was all gone, as were the yaks; every bit of land was now being intensively farmed and the people were again becoming vegetarians. Vast chemical plants were extracting nitrogen from the air, and phosphorus, potassium, magnesium, and deuterium from the sea. Iron and aluminum were being produced in orbit.

Four hundred years after the landing more than a million people were living in orbit, most of them born there. The other planets of the system had been explored but all were either too hot or too cold to be useful and no copper or phosphate rock was found on any of them. It was realized that shortage of phosphorus would limit the population of the planet to five hundred million, and that even at the peak of productivity, which would come about five hundred years after the landing, they would be able to launch only one intersellar ship every five years. Even this rate could not be maintained for long; people in the cities and in orbit were becoming greedy and preoccupied with themselves; the old faith could no longer be maintained. As individual attentions turned more and more to the struggle for personal power and prestige and away from production, the number of space stations would diminish, and though occasional interstellar ships could and would be launched for centuries, the end was in sight.

By now a history of interstellar voyages had accumulated on earth and the planet had it too. According to the latest

77

information eight thousand three hundred twenty seven ships had been launched. Of these, three thousand one hundred and twenty six had broken off communication abruptly for no known reason. Five hundred ten had been lost to mechanical failure so gradual that there had been time to report the nature of the trouble. Six hundred thirty seven others were known to have failed because of social unrest. In these cases the captain had been able to transmit, in a code known only to him, what the trouble was. In most of these instances one or more crew members came to believe that he had personally been visited by God and told that the voyage of this ship was against His will, and that the ship must be destroyed. Thirty nine ships had arrived at stars where there was no suitable planet and no landing was attempted. There had been twenty six landings on small, cold planets where the colonists had survived for a while — in most cases only a few weeks — and then had all died, unable to cope with their new environment. Several landings on planets with favorable environments had lasted more than ten years only to be wiped out by epidemics. Besides this one there had been five enduring colonizations, two on planets of Sirius, one on a planet of Tau Ceti, and two on planets of Altair. Three thousand eight hundred and fifty seven ships were still en route, some of them to stars so distant that the landing would be made by the eleventh generation.

It was believed that the odds were becoming somewhat more favorable. The experience of interstellar travel gained so far was helpful but, more important, earth now had telescopes so powerful that the planets of distant stars could be studied in detail, thus permitting a better selection of targets. The lenses of these new telescopes had diameters of several hundred miles but you could move right through one of these lenses without ever seeing it. They were made of flat mirrors spaced miles apart on a delicate system of girders. An even bigger one was being constructed far out from the sun.

Now that an interstellar ship could be received in orbit about this planet without landing facilities of its own, one had been requested from earth. It was a cargo ship and started with a crew of only six. Even the hull was made of a copper-nickel alloy to be used on arrival. It carried more than ten thousand different kinds of genetic material including bacteria, horses, cows, sheep, dogs, elephants, sea cows, cormorants, and chickens. Plants included a number of one celled types to be used in the sea. To harvest these were shrimplike creatures — some very small — with herring to eat them and tuna to eat the herring. The genetic material required only a little space and weight. Most of the cargo was tungsten, for use in high-speed cutting tools, and cobalt, to be used in these same tools and in computer memories. Sufficient boron, germanium, arsenic, rare earths, and the platinum group of metals to meet electronic needs had been found on the planet. No social trouble was anticipated on this voyage, since arrival was almost certain and comfort and authority would be waiting.

By the time this ship arrived pens and aquaria were ready for the new animals, all of which survived as species though there were many individual deaths. The sea was inoculated with the new plants, which did well there, though their number was soon limited by exhaustion of nutrients. These plants were harvested by shrimplike creatures, processed through herring and tuna, and eaten.

Conditions on earth were known to be deteriorating and it was believed that the social structure there was about to fall apart, leaving the drifting space stations there in rings like those of Saturn.

Four hundred eighty six years after the landing an interstellar ship was launched from this planet in a direction away from earth.

Nothing definite is known about planets of stars other than our sun.

We are situated in the Milky Way which is a nearly circular disc of stars. Its diameter is about six hundred million billion miles. The disc is thicker near its center than at its edges. It contains about one hundred million[1] stars. Looking at the disc toward the flat side we are situated a little off center; looking at it edgewise we are nearly in the center. Most of the stars in the Milky Way are quite like our sun, some a little larger and brighter, many somewhat smaller and fainter. There are a few bizarre types, some blue and very much brighter than our sun, some red and very much larger, but of low density, and some small and glowing white, of a density many times greater than our sun, and some pulsing. These bizarre types probably do not have planets. The space between the stars of the Milky Way contains some gas, mostly hydrogen with a few atoms of heavier elements such as carbon and oxygen, some dust, and fragments of rock or metal known as meteorites.

Peter van de Kamp[2] has given a list of all the known stars which an interstellar ship of the type described here could reach in one hundred seventy years or less. Of these, Alpha Centauri, Sirius, Procyon A, and Altair, are brighter than the sun, and Beta Centauri, Eta Eridani, Tau Ceti, Epsilon Indi, CD–37° 15492, and Omicron Eridani, are more than a tenth as bright as the sun. Tau Ceti, Epsilon Indi, and Altair are single stars and probably have planets; the rest of those more than a tenth as bright as the sun are parts of multiple stars and may have planets.[3]

It is likely that a large proportion of the stars less than a tenth as bright as the sun have planets but that conditions are right for life to have evolved on only a few of these. Faint stars tend to radiate mostly in the far infrared range, which heats any planets they may have but is no use in photosynthesis. Thus a planet of one of these stars which gets enough near infrared and visible light to support life would probably get so much far infrared that it would be too hot for life to exist on it.[4]

Summing up van de Kamp's list of the stars nearest to us, and attributing one habitable planet to each single star more than one tenth as bright as the sun; giving a twenty-five percent chance that any component of a multiple star which is more than a tenth as bright as the sun will have a habitable planet, and a ten percent chance that a star less than a tenth as bright as the sun will have one; we may expect to find about ten stars on which life has probably developed within one hundred seventy years sailing time. Applying this proportion to the whole of the Milky Way,[5] we may expect seventeen million planets on which life may have developed.

Now to compute how many of these planets may be expected to have intelligent life on them. The age of the earth until an expanding sun engulfs it is expected to be about ten billion years. Up to now intelligent life has been present on the earth for about fifty thousand years. If, as suggested in Chapter 2, we literally blow ourselves up within the next hundred years or so that will be the end of it, but if we destroy ourselves by any of the other catastrophes of Chapter 2 we may expect that new and different intelligent life will evolve here within a few million years. Very roughly, we may expect that during its entire history, past and projected, the earth will be inhabited by intelligent life about one percent of the time. Applying this estimate to the seventeen million planets on which life has probably evolved we find one hundred seventy thousand on which there is probably intelligent life now. Continuing the calculation, I define a technological culture as one which has advanced to the point where electricity is used on a large scale. We have had such a culture for fifty years and we expect it to last another two hundred, or about two tenths of a percent of the time we have had intelligent life here. By analogy we may surmise that there are now technological cultures on about three hundred forty planets spread throughout the Milky Way. The nearest of these is probably more than ten thousand years sailing from here. If we knew where it was and beamed a message to it we could not expect a reply in less than two thousand years. Say planet A colonizes planet B, planet B colonizes planet C, and so on, and define a generation of interstellar travel as the colonization of one more planet. What I think the figures of the preceding paragraph mean is that there is a distinct possibility of encountering intelligent life in the first generation of interstellar travel, and that very likely such life will be found by the fourth generation. By the same thinking, it seems unlikely that a technological culture will be found before the twentieth generation but a good chance that one will be encountered before the twenty-fifth. I think these figures also mean that we are likely to be visited by representatives of a technological culture at any time and that there are radio messages from such cultures flowing past us now. Of this more later. That these voyages or messages probably originated thousands of years ago does not seem relevant.

The first plants to grow here, about three billion years ago, used sulfur instead of oxygen in their life processes. For some obscure reason some of these plants evolved a dual capability — they could use either oxygen or sulfur.[6] The dual capability plants were somewhat more efficient than sulfur-only plants and gradually displaced them, and the oxygen they put into the air destroyed the sulfur compounds present. From that time on sulfur-only plants could live only deep in salt water mud, where the oxygen has been exhausted, as they do to this day. The occasional bad smell of tidal flats is due to sulfur compounds these plants put into the atmosphere, as if they were trying to destroy all oxygen and recapture the planet. The dual capability plants, some of which still exist, were for the most part replaced by oxygen-only plants. All animals evolved from these and we have never had an animal which used sulfur in its energy cycle.

The history of other planets may have been different. It is

possible that on some of them efficient sulfur-cycle plants evolved before the dual capability plants and that the latter were never able to take over the atmosphere. It is also possible that sulfur-cycle animals evolved on such planets, and that we may find such as we colonize. It may be that we can colonize a sulfur-atmosphere planet if we already have a colony on an oxygen-atmosphere planet of the same system, but I doubt that it can be done directly from an interstellar ship.

Judging from our experience here many planets should have developed technological culture by now, and if we have a good chance of colonizing planets so had they. Some of them should have made it long before us, say a billion years earlier. Using interstellar ships flying about a tenth the speed of light, and colonizing on the way as suggested here, it would take less than a million years to leapfrog across the Milky Way. Why, then, haven't we seen them?

Perhaps we have. I shall present fragmentary evidence of some cumulative weight to that effect, and one solid fact — that we have found no wreckage of an interstellar ship — which argues against it.

Astronomers have a theory according to which creation is happening now, has always been, and always will be. In this view habitable planets are coming into existence all the time and the age of present planets is randomly distributed, with some very old, some very young and some, like ours, middle-aged. The older planets should have achieved colonization long ago, and if the numbers I have adduced above are anywhere near right, every suitable planet should have been colonized long ago. Perhaps we are a colony — of that more later.

If we assume the "big bang" theory that creation was a single act about twenty billion years ago, the discrepancy between what is suggested here and what we actually see appears to have been avoided since all planets would be about the same age and we just might have been the very first to achieve a technological culture. A closer look spoils the illusion. Even if all the planets were the same age, some of those on which conditions were right to evolve life would be warmer, and some colder, depending on their distances from their star (assuming these to be equally bright; if this assumption is not made the reasoning becomes more complex but the result is the same). There must be a sharply defined temperature at which the development of life is most rapid; above that temperature the process of evolution is slowed down by the spontaneous decomposition of vital chemicals, while below it chemical reactions are too slow for maximum rate of growth. I do not know what this temperature is but it might be about 85° F. Whatever this temperature may be it is preposterous to suppose that the average temperature of this earth is exactly at the optimum. Thus, even if all the planets of all the stars started out together, conditions for the evolution of life must have been better on many of them than they were here, and these should have developed a technological culture before us. Some of them millions of years before us. If that were so one calculates, using the numbers assumed above, that colonizations could vastly exceed the number of planets, which means that every planet on which conditions were suitable for colonization should have been colonized by now, not just once but perhaps several times. Thus, for present purposes, it makes no difference whether we assume the "big bang" or continuous theory of creation.

Implicit in the foregoing argument is the assumption that there is nothing special about our sun, our planet, or ourselves. Any other assumption rules out the use of logic, which is the only vehicle we have for passing from what we know to what we may reasonably surmise. For if we, our planet, or our sun are special, any other star or planet may be special too, and any hypothesis about them is as likely as any other. We cannot gain insight into the cosmos by thinking in special terms.

So we are left with a discrepancy. One way out of it, which will have occurred to my readers, is to assume that I have left out of my calculations some overriding factor which prevents us or anybody else from moving between the stars — some factor which I not only haven't thought about but haven't even guessed. Very likely true. Another possibility is that we have been visited by interstellar ships but not since we have had the scientific know how to recognize them for what they were, and had a language adequate to describe the event in such a way that we could now be sure what was meant. Still another possibility is that some of the "flying saucers" so much in the news in the 1950's were real. It is known that more than 99 percent of these sightings can be accounted for by people mistaking the planet Venus for a space ship, by double exposure of photographic film, by trick phenomena, advertent or inadvertent, by reflections between windows or the optics of binoculars and cameras, by diffraction effects in radar, and by mental unbalance and knavery, but there are a few observations by airplane pilots — practical, well balanced men trained in detecting optical illusions — which have not been satisfactorily explained. It is all very puzzling. My present feeling is that life, and man, evolved independently on this earth but that we have been visited and perhaps colonized since. I present here the fragmentary and very indirect evidence we have to this effect.

The battle between the Gods and the Titans in Greek mythology is clearly the same as that between God and the Devil in Jewish mythology. Something of immense importance did happen here about ten thousand years ago, and we learned something about it and were left a message. The event may have been a civil war within an extraterrestrial culture but what the message is I have no idea. Then we come to Isaiah, who did not know enough about technology to recognize an interstellar ship for what it was, even if he were looking right at one, and whose language was so specialized and nontechnical that we would not now be able to recognize such a ship from his description even if he had given one. He does give us an intriguing clue. He entered a room which was suddenly filled with smoke.[7] Now this is what sometimes happens when one enters a room from an airlock, the momentary drop in pressure causing a mist, but neither Isaiah nor those who recorded and translated his sayings could possibly have known that. He was puzzled by the seraphim who seemed to have wings on their feet and who flew rather than walked. The reader cannot miss the similarity between these creatures and the space men I have imagined inhabiting stations in orbit. It is something to think about. One may wonder also about Elijah's ascent to heaven. The description given in the Bible would fit a modern rocket takeoff.

The Chinese may, just possibly, be the descendants of space castaways. They are a little different physically from the rest of us. Their legends tell us that the first thing their first king taught them was to stand up and walk. This might have been because they had lived weightless and had never walked before. Their language is in a more advanced state of evolution[8] than any other in the world. All other languages of the earth have evolved from extremely complex aggregates of conjugations and declensions and have gradually become simplified, with particles, classifiers, and inference from context replacing complex noun and verb forms. Of the non-Chinese languages (Malay is here considered to be an ancient form of a Chinese unwritten language) English has gone the farthest in this respect, with only vestigial conjugations and declensions left. In Chinese these are absent and probably were never present. Chinese is the end product toward which all other languages are evolving but which none has yet closely approached. It is a highly sophisticated language which may well have come from a culture more advanced than our own. In a colony which almost perished and for generations hung by a thread, the spoken language is the only aspect of the original culture which would survive, and I wonder whether it didn't happen just that way.

The rings of Saturn suggest that someone may have been there long ago. No satisfactory theory which accounts for them has ever been advanced. It is known from radar observations that whatever these rings are made of it is there in fairly large chunks, a yard or more in maximum dimension. They may be the remains of a community of orbiting space stations similar to what is visualized here for orbit around the earth. Abandoned long ago, these may have jostled and bumped together, grinding each other down to torn sheet metal. We know there is no one there now — we would be watching their TV programs if there were.

Any one of these hints that the earth may once have been visited by interstellar travelers would be properly laughed out of court if it stood alone; but cumulatively they carry a certain weight.

If we have been visited by interstellar ships, why has one never crashed and why haven't we found the wreck and fossils of the crew? The absence of any such evidence is a powerful argument against the assumption that we have been visited. The surface of the earth is 70 percent water and the ship would sink. Suppose it were wrecked on the land. Any magnesium or aluminum parts would corrode to white powders which we would not now notice. Iron parts would have rusted away, and copper or brass parts completely dissolved by the rain. Plastic parts would decay into soil under the influence of sunlight. Glass parts would have devitrified and become like stone. Ceramic parts would be so weathered as to look like stone. Even if some parts of power producing, electronic, or navigational equipment survived little-changed, these — because the technology was so different from ours — would not be easily recognized for what they were. One can almost conclude that he could walk, every day of his life, right past the wreckage of an interstellar ship without ever recognizing it for what it was.

Of the structures we have built on earth only the Pyramids, the Great Wall of China, the palace at Escorial, and the Panama Canal would, if they were abandoned today, be easily recognizable ten thousand years from now for what they were, even if no ice age intervenes. Steel reinforcements rust out, cement crumbles under the action of frost, glass devitrifies, and plastics decay. It would be difficult ten thousand years from now to recognize New York as having been a city.

As for fossils of colonists from space we may already have them. The fragments of bone we now attribute to early man could just as well have come from space men.

Although we have seen no men from space, at least in

modern times, perhaps we have heard from them. If so, should we try to get in touch with them? The technical problems involved are considered by Walter Sullivan in his book, *We Are Not Alone.*[9]

When large radar dishes are built far from cities, power-lines, automobile traffic, and manufacturing operations — all of which make radio noise so loud as to paralyse sensitive receivers — and these dishes are turned to any point in the sky, radio emissions are heard. Most of these are random, but some of them cluster about certain specific wavelengths (that is, a radio receiver connected to the dish can tune to them) and some pulsate regularly. The intensity of the radiation varies as the dish is turned toward different points of the sky. Sometimes points of maximum emission correspond to the direction of a visible star, sometimes not, but most of the visible stars radiate weakly, it at all. Our sun radiates very weakly compared with some other stars. Our planet radiates sufficient power from human activities to reach the stars, but this radiation is so diffuse that no radio receiver could be tuned to it, and it is highly improbable that we can be heard on any planet outside the solar system. Points in the sky which radiate strongly but where no star is visible are believed to correspond to dark stars.

Radiation from space to which receivers can be sharply tuned is known to come from atoms and simple chemical compounds between stars and not from the stars themselves. Pulsating radiation from stars is believed to be due to phenomena akin to boiling. The question is whether after the known sources of radiation are taken into account there is anything left.

One is tempted to assume that any radio emission which is repeated faithfully time after time is a message but the question is not that simple. There are theoretical reasons for believing that no signal can ever be faithfully repeated — there is inevitably some garbling. The pulsations of stars are repeated closely, but not exactly, and these are clearly not messages. The difficulty of defining what is and what is not a message is more a problem of words than of substance. If a message comes, and is received and recorded, there will be no practical difficulty in recognizing it for what it is.

And no difficulty in translating it into English either. This, at first, seems impossible. The message may come from beings who do not share our senses of sight, hearing, or touch, and who do not use words. But whoever and whatever they are they must have a number system because without one they could not have developed the technical capability of sending the message. For the same reason they must be acquainted with the chemical elements. Analysis of the light from the stars has shown that the atoms of their chemical elements are exactly the same as those here. We have numbered the different species of atoms of the chemical elements according to the number of units of electric charge they contain, beginning with one for hydrogen through ninety-two for uranium, and on through one hundred-twelve, and perhaps beyond. Any other culture with technological culture as good or better than ours will inevitably number them just that way. It is possible to arrange these numbers in a geometrical pattern in such a way that atoms having similar chemical properties fall into columns and atoms with dissimilar properties into rows. Any culture capable of using chemicals must inevitably have discovered this arrangement. They would therefore recognize the sequence 3, 11, 19, 37, 55 as corresponding to the similar elements lithium, sodium, potassium, rubidium, and cesium and the sequence 3, 4, 5, 6, 7, 8, 9 to the dissimilar elements lithium, beryllium, boron, carbon, nitrogen, oxygen, and fluorine. Thus we have symbols for the English words similar and dissimilar. By further comparing different combinations of atoms we can establish symbols for the question mark (and its inverse, the assertion), for true, for false, for more, for less, and so on. My students have found it easy to decipher messages in codes completely unknown to them. Codes are ordinarily designed to make messages hard to decipher but it is not difficult to design codes so easy to break that they are almost obvious. I do not claim that this is the way messages will be sent and recognized, but it is the easiest way I could think of, and when someone else comes along with a better way that will be all to the good.

A systematic program of listening for radio signals from space was begun at the Green Bank National Observatory in 1959 (Sullivan, loc. cit.) but was abandoned after a few months when no signal was identified. A program sufficiently extensive to determine whether any signals are coming in would cost a great deal and might have to continue for fifty years or more. No private institution could undertake it; this is a matter for governments if it is to be done at all.

The reason why the program would be so expensive derives from the relationship between radio transmitter power, distance reached, and the fineness to which transmitters and receivers can be tuned. The finer the tuning the further a given transmitter will reach. For a message to be recognized and recorded it must come in louder than the random noise from the stars, which sizzle and crackle electrically out there like steaks on a grill. With transmitters and receivers as finely tuned as some we have today a moderate power — say as much as is generated by a conventional power plant — will give an intelligible signal even on very distant stars. Presumably other cultures having technology equivalent to ours can do at least as well in our direction. The transmission may have to be slow by our present standards — perhaps as slow as the equivalent of one English letter per hundred hours. The more finely tuned the transmitter the slower it has to transmit. To find out whether anyone is sending we shall have to have a hundred or more large radar dishes in isolated localities — on the back side of the moon and far out in orbit when we can put them there — each feeding a thousand or more individual radio receivers tuned to different frequencies. The signal from each of these receivers will have to be recorded and compared with the signal received from the same point in the sky and the same frequency at other times. It will be a long time before we can be sure.

Notes

1. You will see estimates of the number of stars in the Milky Way which run as much as a hundred times this. These are based on inferences of stars intrinsically so faint and so

distant that we cannot actually see them. Inclusion of these very faint stars in this calculation would make no difference because very few of them would have the right combination of heat radiation and luminosity to support photosynthesis on any planets they may have, and thus life will not have evolved on these planets.

2. *Popular Astronomy,* Volume XLVIII, Number 5.

3. The question whether multiple stars (some of these, like Castor, have as many as six components) can retain planets is part of the classical "Problem of Three Bodies," which is still unresolved after more than two hundred years of intensive study, and apparently is still beyond the reach of even the most sophisticated computers. The motion of the sun-earth-moon system conforms to the same mathematical equations as those describing the motion of a planet about a double star, and the earth has retained the moon for a long time, implying that components of double stars may have planets revolving about them as individuals. Whether a close-in double star can have planets revolving about the center of gravity of the binary is in doubt. Probably multiple stars are less likely to have planets than single stars.

4. Light in the near infrared has associated with it a characteristic distance only a little greater than that of ordinary visible light. It passes through glass and you can even take pictures by it. Light in the far infrared has characteristic distances two to twenty times that of visible light, does not pass through glass, and affects only specialized military film. Most of the heat we get from the sun comes in the near infrared; most of the heat radiated by the earth is in the far infrared.

5. In the solar system we have one planet, the earth, on which life has developed, and one, Mars, on which it may have developed. By analogy we may expect most single stars more than a tenth as bright as the sun to have one habitable planet and some of them two. Again by analogy, most stars can be expected to have five to ten planets. Some of these will be too hot and some of them too cold for life to have evolved on them. A planet with an average temperature below 20° F. is

probably too cold (if the average temperature is greater than 20° F., the equatorial regions are probably habitable) and a planet with an average temperature of more than 120° F. (one cooler than this probably has habitable polar regions) is probably too hot. These temperature limits apply only to the spontaneous evolution of life (Sagan, *The Cosmic Connection,* loc. cit., would make these limits considerably wider) but planets colder and hotter than this can be colonized from a base on another planet of the same system, by providing mirrors or sunshades to change the temperature as described in Chapter 3.

If the earth were 6 percent nearer the sun its average temperature would be above 120° F. and if it were 5 percent further out its average temperature would be below 20° F. This illustrates the narrow range of distance from its star within which a planet must be for life to evolve on it.

6. The earth's atmosphere originally contained no oxygen but was rich in sulfur and nitrogen compounds.

7. Isaiah VI, 2-4.

8. The statement that Chinese is the most sophisticated language now extant has been questioned but no one has suggested, to me at least, what other language tops it. The question seems to me to be important whether it is relevant to the space program or not. Anything we can find out about the origin of language will throw light on our prehistory.

As to credentials to speculate about this: I read the universal Chinese written language, which no one speaks, and speak Mandarin and Cantonese, which few people write. I am acquainted with the Hunan, Hakka, and Hokien dialects, and have recently been working up the literature on these in the libraries of Southeast Asia. More important for the present purpose, I have a lifetime interest in English slang, especially GI slang, and the way it has evolved during the past fifty years. I think that English slang, especially in its more extreme forms, is a better medium for expressing how one feels than standard English (though this is a better medium for writing maintenance and instruction manuals). Probably it would be better if we did as the Chinese do and accept slang, considered as a dialect, as our spoken language while preserving standard English as a written language which no one speaks. As far as our understanding of the cosmos goes, the important point seems to be this: every change in English slang during the past fifty years has taken it further away from standard English and closer to Chinese syntax.

Chinese has two parts of speech, the classifier and the particle, which are not ordinarily recognized in English grammar. In Chinese it is indecent to let a noun stand alone — it must be heralded by a classifier. Not a man but "one piece man"; not a letter but "one envelope letter". I remember sitting in church in China and listening to a sermon, not doing very well with it, but I knew Father was expounding on ten strings of something which later turned out to be "ten strings commandment". Now your GI has this same horror of naked nouns; he supplies his own classifiers, most of them of pornographic origin. Then there are the particles. Confucius used more than thirty of them, all of which can be best translated as "huh" insofar as they can be translated at all. The GI uses these, as the Chinese do, to express tense. "I see him on the corner, huh?" instead of "I saw him on the corner."

It may be that Chinese is sometimes considered a primitive language because it makes less use of human ability to produce and distinguish sounds than English does. If, as most scholars do, we consider Chinese to be a monosyllabic language, there are only about four hundred distinct word sounds, and of these at least two hundred are so similar that they cannot easily be distinguished except by a native-born Chinese. In the written language this causes no difficulty because, though there may be several hundred words which are pronounced the same, the characters for them are different and easily distinguished. In the spoken languages the difficulty is partially overcome by enunciating words of the same pronunciation but different meaning in different tones, and so distinguishing between them; but this prevents the tone of voice from being used, as we use it, to convey shades of meaning or feeling. Thus, while written Chinese is the most sophisticated language now being used, spoken Chinese is one of the most primitive.

The paradox is resolved if we think of the Chinese as a highly sophisticated culture who went through a shipwreck which was very nearly the end of them. They preserved syntax through many generations of bare survival but during this period their spoken language deteriorated almost to the level of sounds made by animals.

The reader will have noticed a difficulty with this theory of the origin of the Chinese. Our other races can interbreed with them. This seems to imply that either they originated on earth, developed an electromechanical capability, went off into space, became highly sophisticated there, and then returned to earth, as some animals have left the sea and returned to it, or that some members of the human race were carried off by a higher culture and then returned here, or that the total number of feasible evolutionary paths from a chemical to an electromechanical culture is so restricted that any who achieve it must be very like ourselves. I am inclined toward this latter hypothesis.

9. New York: McGraw-Hill, 1974.

7.   Men versus Machines

Well, we have proved our point—we have gone into space and survived there — live people just like us have walked on the moon and returned to resume the activities of live people just like us. Why not sit back and let machines do the rest of it? They have been further out and into more bizarre environments — near the sun to photograph the planet Mercury, inside the thick clouds of Venus, past that object of our insatiable curiosity, Mars, and on to Jupiter and beyond — than we have, and they have done these things at a cost less than manned ships could have done them. We have learned at least as much about the nature of the solar system from machines as we have from our own trips to the moon and our study of the sun from Skylab. The answer is that for us to bow out now in favor of machines would break the chain of the evolutionary process and thus, by *force majeur*, we will not do it. I think this is the right answer, but baldly stated it is too pat; we must inevitably be suspicious of it and dig to see what is behind it.

The evolutionary process which has brought us from the ocean slimes to where we are now has been primarily chemical, though electrical effects have probably been important from the start — chemical potentials across cell membranes, chemical batteries in the nerves — and perhaps in other ways.[1] The system has evolved in such a way as to produce controlled mechanical forces as well. Thus we arrived about a million years ago at a point where we had, in addition to an extraordinarily effective chemical system, an electromechanical system too; but that was a dead end.[2] There was no direction in which this system could evolve further in such a way as to lift us out of the atmosphere, though some of our feathered relatives have explored fairly high up into it. Then by infinite trial and error a new way was found. The property of imagination was developed by the classical process; imaginative mates, at least in the male line, did better in competition than dull mates, and we managed to breed the quality of imagination into the human race. From that came science and engineering, and we got around the dead end by learning to produce machines which could and did lift us out of the atmosphere. We could not have done it without machines and they could not have done it without us to build them. But these machines were simply larger and stronger versions of ourselves; they used the same electrical control and mechanical force that we do. The only things that were basically different were that they could not think and could not reproduce. They required contact with a human intelligence at all times in order to function.

There is room to quibble here. We operate at nearly constant temperature. The greatest difference in temperature which we can tolerate inside our vital organs is less than 20° F. or less than 4 percent of our working "absolute"[3] temperature. In a rocket ready for takeoff the corresponding difference in temperature between different parts is about 80 percent, and in the same rocket just after takeoff it is about ten fold. It may be stretching language to assert that the difference in chemistry between our bodies and that of rockets is merely one of degree rather than kind. Further, though we use electricity in controlling our bodies there is nothing in the body analogous to electric motors. Present rockets have a few such motors. The power generated in them is less than one part in a thousand of the power generated chemically in the rocket engine but they are there, and their presence may logically destroy our analogy. These questions I leave to expert semanticists. They are moot anyway because we are about to have nuclear rockets, and by no stretch of the imagination can it be asserted that these function in a way analogous to our own chemical production of mechanical force.[4]

As we send machines deeper and deeper into space it will

become harder and harder for human operators to furnish adequate control. The difficulty is fundamental. There is necessarily a time interval between an observation by a machine and instructions from an earthbound operator what to do about it. For machines on the moon this time lag is about three seconds. It would have been impossible for an earthbound operator to have controlled the cars we saw running around on the moon. By the time an earth-based operator became aware that the car was about to hit a rock it would have been too late to prevent a crash.[5] The television cameras on the moon were controlled by earth-based operators and we saw that their performance was sluggish and inadequate. On Mars the time lag will be about ten minutes — too long for an earth-based operator to be of assistance in a crisis. As our vehicles flew by Jupiter the time lag was about an hour. These machines succeeded only because they had built into them an intelligence comparable with that of an insect — they could act in response to their own observations of local conditions without waiting for instructions from their human operators. At the nearest star the time lag will be about nine years.

The further out we send our machines and the more complex the missions we assign to them, the more intelligence we will have to build into them. For the present this does not matter because the cost of the intelligence we build in is only a small fraction of the total cost of the mission, and if we double the intelligence, or even increase it ten fold, it will have no substantial effect on the total cost; but this will not be true much longer. A machine intelligence able to accomplish what we expect from our manned mission to Mars may well involve a weight, bulk, and cost[6] comparable to that of all the rest of the mission.

It is possible now to build machines which can think (present-day computers don't, as anyone who has tried to correspond with the Internal Revenue Service will attest) and the only reason we do not do so is that such machines would cost more than we now pay the human intelligence with which they must compete. This may not be true very much longer. While the cost of almost everything else goes up and up the cost of computer parts is dropping fast, and there are indications that it may go very much lower.

For a while at least, machines, if they are to think, must think logically and this may prove a severe limitation on them. The human intelligence succeeds precisely because it is illogical. From illogic comes the ability to improvise, to compromise, and to dare. We all know individuals who are highly intelligent and logical but who have failed in their personal lives, and others, less intelligent and logical, who have been brilliant successes.

If we thought logically we might soon die out altogether. As individuals we do not need children, as our ancestors did, to care for us in our old age. Our pensions take care of that. It costs a great deal to raise a child, and when we have done so he is, more often than not, a sorrow and a disappointment to us; but illogical couples go right on having children, and that is what keeps us going.

I have no doubt that machines with intelligence which can improvise and dare can be built — even machines which can maintain themselves and reproduce, and can originate missions and evolve as we do — but I wonder how big these brains would have to be and how much they would cost. In other words, how practical would they be as colonists of the nearest stars? I can imagine them making the first colonization, but how they could build up an industrial civilization, starting with practically nothing, and then go on to send out interstellar ships of their own defeats me. I have tried to make the calculation but it is too complex for me; I can reach no conclusions, and I even wonder whether the choice between men and machines to carry on the evolutionary process is really one of cost.

I must now come out explicitly with the premise which underlies this entire line of thought. I know it is indecent to say such things because we have been brought up to think of human life as sacrosanct[7] and valuable beyond all calculation, but the fact is that human life is the only resource we have which is free, and the only one we have which we can squander without ever running short of it. I think this is the reason why we, and not machines, are destined to carry through the evolutionary process.

Notes

1. In some fresh water eels the electric generating function has usurped the control function and they can give formidable shocks.

2. Nature has even experimented with jet propulsion. The bombardier beetle generates inside his body two sets of chemicals, one of them hydrogen peroxide (which we used in our first rockets), and the other hydroquinone. These he can mix suddenly in a special chamber where they flash to form a violent jet. Like us, he now uses his rocket as a weapon.

3. Absolute temperature is reckoned by adding 460° to the Fahrenheit temperature. We cannot use Fahrenheit temperature for figuring percentages because that scale of temperature has a zero. The absolute temperature is so defined that no real object can have a temperature of zero. In deep space an object which generates no heat gets down to about 15° absolute.

4. It is true that for millions of years our bodies have contained the radioactive element potassium, and that the radioactive decay of potassium inside us does contribute to keeping us warm, though the contribution is extraordinarily minute. The presence of potassium in our bodies may have had a decisive effect on our evolution, but to assert an analogy between the decay of potassium in our bodies and the function of a nuclear rocket would be too far-fetched.

5. The Russians put a vehicle on the moon which was controlled by an earth-based operator. Because of the time lag this vehicle could move only very slowly and apparently it accomplished nothing.

6. In reckoning comparative costs between manned and unmanned missions we must not forget that unmanned missions are one-way whereas manned missions are two-way (travel to the stars is an exception because two-way missions that far are impossible). Two-way missions are very much

more expensive than one-way missions — not just double the cost. Further, if a machine gets in trouble it can be junked without kicking up a scandal.

7. The concept of simultaneous infinite and zero value on human life is not so radical as it first appears. It is a truism of philosophy and science that opposites, though obviously such under conditions of low energy or stress, become more difficult to distinguish at high stress, and become indistinguishable in the limit of infinite stress. Love and hate are easily distinguished at low intensities but become identical at high intensities. Positive and negative electrons are different things at low and intermediate energies but become more and more alike as we go to higher energies. The same is true of photons and neutrons. Under the stress of modern life there is nothing inconsistent in our thinking if we assign, in different contexts, zero and infinite values to human life. We may think of these as two equally valid solutions of the same equation, one or the other to be used according to the nature of the problem we are working on. They are really the same solution. We should continue, as we do now, to spend large sums to save single individuals in danger, as from shipwreck or mountaineering accidents, and we should also continue, as we do now, to let millions starve rather than get bogged down ourselves in social problems which we cannot solve.

Whether we commit ourselves heavily to the space program or not we can accomplish nothing until we get over being squeamish about casualties. In the whole of history nothing of consequence has ever been accomplished without heavy casualties to men and institutions, and there is no reason to suppose that things are about to become different. We are gamblers by nature. We would rather play a long shot with a chance of a decisive improvement in our life style than to make an almost sure bet on a marginal improvement. We buy our lottery tickets; we may die early in dreadful circumstances or much later in luxury, as things turn out.

We philosophers may assign zero or infinite value to human life, but there are others who, quite rightly, assign a cash value to it. Judges and jurors do this every day; legislators and police do it as they write traffic rules. Use almost any set of statistics you like — deaths by automobile accidents, fatal accidents in the construction industry, and so forth, and you always come up with the same answer. A man's life is worth what he can expect to earn in his entire lifetime — on what he as an individual, taken with all his personal strengths and weaknesses, can expect to earn, not the average of what other people earn.

Probably the greatest single criticism which can be leveled against the United States Atomic Energy Commission is that there have been so few fatalities in connection with its operations. Too much has been spent on safety; too much attention has been paid to ill-informed quibbles by idealists. If the Commission had gone ahead with the things that had to be done, placing a fair but not exorbitant value on human life, we should now have the nuclear power plants we need to tide us over the present energy crisis.

8. The Short Term

We can see how the space program will go in the long term once we recognize that it partakes of the inexorability of the evolutionary process, of which it is an essential part. Things will happen, not because anyone plans them that way, or because anyone wants them to, but because they must. When you see it this way it is pointless to inquire about the path by which the result comes about. The only things that matter are weights, times, distances and energy, and the chemistry, biology and physical characteristics of distant planets. In the short term, though, we have to go into social and administrative problems, and into questions of money and other resources.

Before we decide where we are going, we need as good a navigational fix as we can get on where we are. The problems we are now having — inflation, shortage of energy, and terrorism — are the birth pangs of the space age; but knowing that doesn't ease them. The fact is that a phenomenon utterly new in human experience has sneaked up on us unawares. This is the obsolescence of skills. A man can no longer aspire to be what his father was or to live where his father lived. Virtually all the technological skills a man has learned by the time he is twenty becomes obsolete before he is forty. Whole professions which have been with us for a thousand years may disappear in ten. Domestic service is gone. Clerks are about to disappear. Carpenters are going the way of our forests. Repairmen have gone. It is cheaper to buy a new machine than to fix an old one. We have now to rethink our goals, to bring them into line with what is possible, and rethink our institutions before they fall apart.

First, education: Our present system of public education[1] is built on the premise that we train the boy to do what we expect the man to do. Right now we haven't the foggiest idea what the man will be doing by the time he is fifty, and it is absurd to pretend to teach him to do it. Everyone who has the capability and wishes to remain a viable member of the industrial community now has to be retrained every few years so that he can do his job during the next few years. Most of us will be spending some time in classrooms all the rest of our lives.

The mission of our public high schools and colleges has been gone for twenty years but they still go through the old motions. Actually they are now little more than child-sitting facilities, keeping the students out of their parents' hair, and out of competition with organized labor as long as possible. Even if we could afford it this is undesirable, and anyway we can't afford it. Education is the only soft spot left in federal, state, and municipal budgets, and if for no other reason it will get a searching look. If, as I assume here, the rethought purpose of education is to focus narrowly on the object of retrainability, we could have better education than we have now for less than half what we now pay.

There should be no compulsory education beyond the eighth grade, and no student should be allowed[2] to go further in public institutions unless he demonstrates outstanding ability and a real vocation for learning. Most of those finishing the eighth grade should go directly into factories or, more likely, construction work.

Many of our ideas about education and the place of children in the national economy derive from a period which ended about fifty years ago, in which children were viciously exploited as cheap labor in factories. These abuses were real and shocking, and there is no thought here of repeating them. We have learned enough since then about how to regulate industry to insure that this does not happen. On the positive side, I know from my own experience that the best way to learn is by actually doing. Making things which will actually be sold and used, or building structures one will see for the rest of one's life is incomparably more character building and maturing than the meaningless gabble now heard in our high school and college classrooms. Not that people under twenty should mingle with and become indistinguishable from the rest of the work force. They should spend at least four hours a week in classrooms where they should be taught the engineering reasons why they do things the way they are told to do. They, with all the rest of the work force, should be invited to attend free evening classes in which they can explore as far as they like into history, literature, art, music, and science.

Junior workers should be paid no more than half as much as their senior colleagues. The rest of what they earn should go toward providing voluntary lifelong classroom instruction to those who want it.

Organized labor will no doubt object to competition by youngsters, but organized labor, now fighting for its life, has more important things to worry about and may not give us much of a tussle on this one.

Getting back for a moment to primary school instruction, we shall be asking it to do everything worthwhile which high school and college now do. The students are more than able to absorb this at that age; it is the teachers who will be bottlenecks. We can relieve them of much of the drudgery. We can forget addition and multiplication tables and redefine the operations of arithmetic in terms of the pocket calculator. This works even better as a base for algebra and calculus than the older definitions. We need no longer try to correct slang, which will be the working language of the students, and will teach standard English as a foreign language. We should leave history and literature and the mechanics of government to a time when the student is mature enough to understand them, when and if he wants to know more about them. Art and music should probably be retained in the elementary cirriculum to give the children some opportunity for self expression. We can give up the "ain't science wonderful" bit, with its mystery, swirling electrons, and theories of the nature of matter, and confer instead a solid appreciation of the significance of weight, time, distance, momentum, force, and energy. Practical aspects of chemistry and physiology should also be taught. Most important, the objective of all this must be kept clearly in sight; curriculum and instruction should focus narrowly on the single goal of retrainability, so that as the student becomes adult he will be agile, able to adjust rapidly and effectively to the changing and unpredictable situations he will encounter as an individual in the space age. The emphasis must be on imparting an exact understanding of written and spoken standard English; he must be able to understand exactly what is meant, not approximately what is meant, by everything he reads and hears in that language. This will require endless drill. Students don't like that, and teachers don't like that, but they will just have to do it.

Besides understanding exactly what is said to him, he needs experience which enables him to think logically. Elsewhere in this book I have expressed misgivings as to the reliability of logic as applied to basic problems of philosophy, but it is a great help in practical everyday decision making. We have to learn to cut through floods of words and endless side issues and get through to the heart of a problem. I do not believe that accurate thinking can be formally taught, but I do know of two ways in which the necessary practice and experience in the use of logic are accessible to young children. One is the study of geometry and the other is playing chess.

Geometry is an indispensable tool in engineering, but this is a byproduct rather than the object of teaching it. It should be taught as a drill in passing from facts to consequences. Mixing geometry with arithmetic as we do now in our elementary school curricula takes the bloom off of both of them.[3]

Chess teaches how to face up to unpleasant facts and do the best one can in the situation actually existing, rather than wishing things were different.

Our blue collar population is probably about to disappear as such within the next decade, and certainly within the next two, splitting into two fractions: one composed of the agile and adaptable, willing and able to be retrained every few years for different jobs, and willing to move, with their families, to wherever work is, and the other composed of those of any age who are unwilling or unable to be retrained. These will be put out to pasture on subsistence farms in the Middle West (there will be no room for them in the coastal cities). The agile and adaptable will run huge construction and earth-moving machines and will be highly paid, as airline pilots are now. Within the next decade manufacturing, transportation, and retail sales will be almost completely automated. Present day blue collar workers who think they are striking are, in many cases, actually abandoning jobs which can be done better by machines.

We are committed, rightly I think, to the proposition that no one in the United States need be undernourished. It makes no difference whether one can work or can't, or whether one is deserving or not. We are also committed to furnish service-able clothes, a warm and dry place to live, and medical care. I think this is as far as government should go. We are not com-mitted to furnish a car, a television set, a home in the place where one grew up, or enough cash to shop in a supermarket. We need to move people on welfare out of the cities, particu-larly those on the coast.

Old people like me constitute a baffling problem. Grandma no longer has the power she once did to disrupt a young household, but we are still far from escaping the tyranny of the aged. Until we can think of something better I suggest, as a working rule, that we old people be allowed to make what use we choose of our Social Security benefits and such other money as we have been able to amass for our retirement, or can earn, to live where and how we wish, but we should re-member that we are here on sufferance, and mind our manners. Government should treat old people who for any reason have failed to finance their retirement like any other indigents. It should make no difference whether they failed through improvidence or bad luck, or even that they gave away what they earned to help others. Private charity may take a different view, singling out deserving people like me for special privileges.[4]

This may be the place to jettison a burden of hypocrisy so heavy that it could still drag us down. We say that we want everyone to live about as well as everybody else. We do not go quite so far as to say "from each according to his ability; to each according to his needs" but we feel we ought to subscribe to something like that. We feel that living better than others because we have more money is slightly indecent. Actually very few people in the United States want it that way. As was mentioned in Chapter 7, most of us are gamblers at heart and want to play long odds. Even the poorest of us would rather have a chance of a great improvement in his life style than a near certainty of a marginal improvement. Chance for great improvement in our individual fortunes can come only if those who have money live not only better, but very much better, than those who have little money. If we are frank with ourselves we admit this is true, and if we could bring ourselves to admit it freely and without any sense of shame our decision making procedures would work a lot better.

Now, what do we really want of the future? I think that what we want most is social continuity, a condition in which we can plan our personal lives with a good expectation that things will work out much as we plan them. We can stand rapid social change if it is orderly change and predictable change. What we cannot stand is improvision — uncoordinated stop-gap measures designed only to meet immediate crises. We cannot stand many more rude shocks like the quadrupling of oil prices and the sudden rise of food prices either. These were induced from outside our country. We must recapture our own destiny by becoming again an oil-exporting nation and building up reserves of grain and soybeans which will insulate us from disasters elsewhere in the world. All these things we must do in the short term, no matter how we go about the space program.

Within the above limitations imposed by our present situation we can now begin to map the paths which we can follow forward. Almost every path we could conceivably take is contained within one of three options: we could devote all the strength we have left after setting our house in order to humanitarian goals, carrying the space program at a level of expenditure lower than we do now. We could defer decision, devoting our entire effort for the next decade to solving the problem of energy and inflation, carrying the space program at about the same level of expenditure as we do now. Or we could elect now to make the space program the central feature of our planning, and build our measures for beating inflation and energy shortage around it.

The argument for the first option is that while there are people starving, and while there are criminals to be rehabilitated and restored to society, and while there are youngsters getting less than sixteen years schooling, almost all the funds which might be allocated to the space program should be spent in these areas.

There are people starving, and even the most narrowly educated scientists know this, and are distressed by it, and would do something about it if they could. But those of us who have looked into the matter know that the problem is political, not technical, and quite insoluble. The United States could, without diminishing our diets in any way, or requiring any substantial sacrifice on our part, produce and deliver enough food wherever it is needed in the world to bring the diet of every human being to a level adequate for good health if Americans were allowed actually to hand it to the individuals who need it. If we turn it over to local governments for distribution the officials either steal it or divert it to propaganda purposes. To prevent starvation we would have to actually take over the government of the countries concerned. After the way we have just been burned in Southeast Asia there is no chance that we will consent to take over any other government, so we cannot prevent starvation, no matter how much or how little we spend on the space program.

The problem goes deeper than that. Outside of the Anglo-West European community, in which no one is starving, few are concerned with the starvation[5] of others, and some actively desire it. As I have related in Chapter 2, I have seen starvation in Calcutta at a time when there were stores of food rotting just outside the city. Thirty years later India spent its meager resources of hard currency to develop an atomic bomb while people starved for lack of food which that money could have bought. Nothing has changed. It is no use trying to help people like that. Most of the African and some of the South American countries are just as bad as India in that respect. Those nations, including Russia and China, who make the most effective use of the food which we send them pay cash for it.

For many years, at the behest of humanitarians, we have been spending more and more on our penal system and the rehabilitation of criminals, and all the while the crime rate continues to rise, as of course it will until the money spent in those areas is sharply curtailed. The experiments have been done, the results are in, and we know now that no amount of effort will save a man from himself. It is time we cut our losses and got out of an unprofitable area.

For many years it was supposed that most human shortcomings were due to lack of understanding and that the way to kinder and more harmonious human relationships was more schooling. Tens of thousands of highly intelligent people have devoted their lives to demonstrating this principle, and money equal to several times our annual budget has been poured into their effort. It was a good and necessary experiment, and now we have the results. We have learned that schooling beyond eight years, plus a little instruction in technical skills, has little to do with a man's prospects for a satisfactory[6] life or his ability to live harmoniously with others. In Northern Ireland, where most people are schooled well beyond the eighth grade, a senseless war grinds on year after year while in Malaya, where the general level of schooling is well below this, and where ethnic, religious, and cultural differences are greater than they are in Ireland, the same types of problems are being solved peaceably.

We also know that there are some of us who enjoy music, art, literature, and history and many of us who are bored sick with these things and derive no profit from them. This is not said critically. I do not want a society in which people are cast into a common mold because I do not wish to be molded myself. The point I am making is that it is silly to spend money to force people to learn things of no interest or use to

112

them. We should cut back our budget for education and make schooling beyond the eighth grade so expensive that only those who can truly profit from it will bother with it or be helped to obtain it.

The British Indians are, I believe, more intelligent than we, at least in the sense of being able to manipulate abstract concepts better, and they are certainly better philosophers — in fact, they never miss a chance to assert moral ascendency over us. The higher echelon officials there are, on the whole, better educated than ours, yet India is among the worst governed countries in the world and it is the place where an individual has, at birth, the least chance for a satisfactory life. To me this proves a great deal. It reinforces my conviction that we should spend less on schools than we do now, and that the call for more money for education is not a valid argument against a large scale space program.

The argument we can advance to humanitarians in favor of spending more on the space program is that we now have enough material things, except for a temporary shortage of oil, and that what we want now is stature, in the form of jobs which are not routine, which tax individual intellectual capacities to the limit, and of which the individual can be proud. This can be accomplished only by the automation[7] of routine jobs. If we do this our production per man will increase to the point where it cannot be absorbed in better living. There are only two places where this increased production can go. One is into munitions and the other into the space program. The space program seems preferable.

In the United States we already have more to eat than is good for us, more clothes than we have closet space for, and most of us have adequate houses. Our automobiles, television sets, and other entertainment facilities are more than adequate, but most of us are working at jobs we hate and which diminish our dignity as human beings. Working on an assembly line, making automobiles, most of which will go to families which already have one, is not a job we can be proud of. In a fully automated plant no one will do anything until the machinery fails at some point; then the person in charge will

113

have to think clearly, accurately, and fast to correct the difficulty. The job will be a succession of crises, all different, and each taxing the worker's ability and ingenuity to the utmost. These will be days to be proudly remembered and to brag about.

At this point we have to face up to an ugly social problem which is right on top of us and which we shall have to deal with whether we spend much or little on the space program. As our society becomes more complex so do our social and administrative problems, and an increasing fraction of us, for lack of sufficient intelligence and self discipline, are unable to cope. No amount of schooling, counseling, or psychiatric treatment helps; a person in this situation rarely gets out of it and cannot aspire to a meaningful job. This has to be accepted as a fact of modern life. Those of us who can cope are not going back to a pastoral life for the benefit of those who can't. These unfortunates can be employed in forms of agriculture which are difficult to automate, in the endless sorting required in recycling and cleaning up the environment, and in janitorial and domestic service. A constitutional amendment will be required to regulate their status.

The nature of our society is changing. A family whose background is in science and engineering now has a decided economic advantage over a family whose background is in the liberal arts. Honest and capable repairmen — auto mechanics, television servicemen, plumbers, carpenters, electricians, and the like — are vanishing from the scene, and those who remain insist on being paid as well as trained professionals and charge accordingly. In an engineering oriented family when, say, a lampcord deteriorates or a faucet leaks, the man or woman of the house just fixes it and forgets it; for a liberal arts oriented family a repairman must be called; this puts a strain on an already precarious budget and the bungled repair puts the coup de grace to an already exhausted patience. The engineering oriented family may elect to send the car or television set to a repair shop, but they can at least tell whether the repair has been properly done, while the liberal arts oriented family has doubts whether anything was

114

done at all — doubts promptly confirmed when the old difficulty reappears in aggravated form — for the repair shop people know whom they are dealing with. They have to do better for the knowledgeable customer than for the less well informed. Or if a new closet or bookshelf is needed, in the engineering oriented home the man does some measurements, makes some drawings, works out a schedule of materials, buys these, and he and his wife install it. They get exactly what they expected. In the liberal arts oriented family a carpenter must be consulted. There is an uneasy feeling that there has been no meeting of the minds, and indeed the result is not exactly what was expected and desired. To cap it all, the liberal arts oriented family pays three times as much for the new closet as their engineering trained neighbors would have had to.

Much the same advantage accrues to engineering oriented families in financial management. Nowadays, when we write a letter to the credit department of a store, or to the issuer of our credit card, or to the Internal Revenue Service, or when we try to extract some of the benefits due us under our health insurance plans, we are not writing to a person but to a computer, and the letter we get in reply is not likely to be helpful or informative. Those of us who understand what goes churning around in the instruction files and memory banks of a computer are merely saddened by this; those of us who do not understand the details of how a computer works, and who cannot get rid of the idea that computers should act something like intelligent human beings, are enraged, with damage to our ability to cope and to collect what is due us. Computers are simple-minded things — it is easy to diddle[8] them if you know how they work — and it is fun to write a letter which will make them blow their fuses and call in human aid.

A result of this sort of thing is that liberal arts people are being pinched out of the comfortable layers of the social order and replaced by engineering people. Thus the nature of our group is changing in such a way that we are less impressed by humanitarian arguments and more anxious to engrave our mark on the physical world.

All right, but what about rugged individualists in the

liberal arts who refuse to be pinched out and become do-it-yourselfers as a matter of preservation? This can be easily accomplished and is fun. The path is different but the result is much the same as a transfer to the engineering branch. For to do so simple a thing as replace a worn electric lamp cord one must give some thought to what the cord does, how it does it, and why it is constructed the way it is. Such cogitation leads into realms of thought quite as fascinating as those of art and literature, and exploration of these tends to divert one's interest away from questions of what should be done socially toward questions of what can be done physically.

Teenage students who have not yet decided on a career, who are intelligent to a degree which those of us who watch their antics find difficult to accept, and very much aware of the importance of money and the esteem of their peers, have come to realize that only an extremely gifted individual has more than a remote chance of impact in the arts or literature and of enjoying the perquisites therefrom. They know, too, that less gifted individuals can do sound, useful, and even brilliantly imaginative work in engineering design, and that the rewards for such work, in money and prestige, are quite satisfactory. They cannot know, though it is just as true, that having written a book no one will publish, or done paintings which he cannot sell, or written music no one plays, can be a soul-scarring experience, while designing a perfectly ordinary machine, following it through preliminary sketches and calculations through working drawings, watching the individual parts being fabricated in the machine shop, watching these being put together on the assembly floor, and then seeing the machine actually do what it is designed to do, is an immensely satisfying emotional experience. Realization of such facts is shifting college enrollments toward the sciences and away from the liberal arts. All this is cumulative; the nature of our society is changing and the effect will be less interest in humanitarian problems and more in physical.

Taking all these things into account I doubt we shall choose the first option, that of going all out to solve our social problems while downgrading the space program.

116

We may very well choose the second option, that is, to get our house in order by beating inflation, energy shortages, and high food prices, and bringing our educational, labor-management, and welfare systems into accord with the facts of modern life, while carrying the space program at a low level and deferring our ultimate decision as to what to do about it. The space program itself might profit by this. If we put money into it too fast we may get such a large investment in specific devices and plans that we cannot abandon them when they are found to be defective. We may already have done this in connection with the shuttle program (see appendix 5). The best engineering comes from austere rather than plush establishments. But we must not delay our decision too long. If we wait until we are fully prepared, socially and economically, to cope with the space program, we shall never get to it. As every chess player and military man knows, if you wait until you are fully ready before you attack, you never will, because the other fellow will attack you first.

Let us look at some reasons why we may choose the third option, which is to press on with the space program as rapidly as our resources permit. Probably the most compelling of these is, first, to do something about the weather, along the lines indicated in appendix 6. The price tag is well within our means. Second, once we can operate from space we can have all the clean solar energy we want at prices competitive with nuclear power plants. This subject is treated more fully in appendix 4. Third, there are some manufacturing operations which can be done cheaper in space than on the ground. These include growth of single crystals for use in electronic devices and in solar energy conversion, and the fractionation of biological materials,[9] which may provide new and powerful medical aids. Fourth, to satisfy our intellectual curiosity. Except to mention the manned landing on Mars and the possibility of having optical and radio telescopes on the side of the moon which we never see, I shall not enlarge on this point which has been so adequately treated by Carl Sagan in *The Cosmic Connection* (loc. cit.). Fifth, to provide room for expansion for communities who have run out of land area where they are. Hong Kong may have the first wholly owned space suburb and manufacturing complex! Sixth, a sense of destiny which impells us to provide a refuge for some of our descendants if the earth becomes uninhabitable. Seventh, what we shall have to do to get the space program moving will solve some of our present problems and give direction to our efforts to beat inflation and solve the energy crisis.

If we do go ahead on a larger scale with the space program we need first to start whatever programs are needed to insure that we have enough power available by the time it is needed. Practically this means that we must begin at once to construct many new nuclear power plants and build facilities for converting coal and oil shale into petroleum-like products, including the hydrocarbon fuel which we use in the booster stage of our rockets. Details of how this can be done will be found in appendix 3. I need only note here that by the time we are ready to put thousands of people in space we shall have all the electric power and gasoline we need for industrial and domestic use, and will have become an oil exporting nation.

For 1975 we have scheduled a joint exercise with the Russians. This is essentially a docking and crew-interchange exercise which will give both nations the capability for rescue operations when a ship of one country or the other gets into trouble. I think this program is of great importance. It may even avert a nuclear war, but in itself it is a dead end. It is only preparation for the bigger things to come when we have a working shuttle service.

By a shuttle service I mean a fleet of completely reusable vehicles plying back and forth between earth and orbit and carrying material into orbit at a cost of less than three dollars a pound. Except for a few shuttle buses, which will carry men to and from orbit, probably at a cost of about ten dollars a pound, these will be unmanned vehicles, and will in fact get so hot inside during the returns to earth that no man could survive in them. The National Aeronautics and Space Authority's present plan will not produce this kind of vehicle. They visualize a manned, only partly reusable vehicle, which will not be economical. To use it to deliver material in orbit will cost

118

more than a hundred dollars a pound. The reason for this, with details of shuttle design, are given in appendix 5.

As soon as a working shuttle service is in operation we can begin to put large numbers of men up there. Their first task will be to build barracks, as described in Chapter 3. Then they will build hangars and manufacturing facilities. Some of them will then begin to make and launch the mirrors described in appendix 6. Some of them will begin work on the space-based system for converting solar energy into electric power and beaming it to earth as described in appendix 5. Some of them will manufacture crystals and biological materials for sale on earth. And some of them will manufacture and assemble material for the manned landing on Mars.[10]

Unlike the moon landings this will not be a gallant dash against the odds, carried out by men as skillful and daring as the race has yet produced, but which would have failed even so except for incredible good luck. The voyage to Mars, when it comes, will be a methodical effort made by large numbers of quite ordinary people, and the actual landing, when it comes, will be almost casual.

The moon landings could have been like that; perhaps they should have been, but if we had gone that course we would still be five years away from touchdown. Instead of dashing to the moon we could have put our resources into building a shuttle service, established a continuously manned colony in orbit about the earth, built a nuclear tug there, towed several of the barracks to orbit around the moon, left these unmanned at first until we had provisioned them, then manned them, received shuttle vehicles to take us safely down to the moon and back, and finally made the trip to the moon's surface as part of the day's work.

Some useful parallels can be drawn with our conduct of the war against Japan. At Gaudalcanal we gambled against the odds, had more than our share of luck, just squeezed through — and gambled no more. For a type of management which can make a nice calculation of risk, can act boldly, improvise brilliantly, produce heroes and supermen as needed, and charge right up to the edge of disaster without slipping over it, is also a type of management which knows enough to quit when it is ahead. After Guadalcanal we never risked defeat; we bulldozed our way across the islands of the Pacific, and finally walked into Japan almost casually, unopposed.[11] This is why I think we shall use the methodical approach to the Mars landing, even though doing it that way may delay it by as much as ten years.

If we proceed to Mars methodically as outlined above, we shall see, not long after the shuttle service comes into operation, two new types of space hardware — the nuclear rocket and the cold weather living quarters which will be needed in orbit about Mars and further out. The nuclear rocket will be small, unlike its chemical opposite; it will generate no noise and only a slight flash. Its thrust will be as gentle as that of your car pulling away from the curb, but it will be able to maintain that thrust for weeks at a time instead of minutes, as the chemical rocket does. Because of its low thrust the nuclear rocket cannot be launched from the earth; it must be built in orbit. The cold weather housing will be a space station, much like those described in Chapter 3 and appendix 1, but with glass ports over almost its entire surface with the center of the vehicle located at the focus of a large parabolic mirror. Behind the mirror there will be a cylinder of blackened surface for re-radiating the heat collected by the mirror. Appropriately, the whole assembly will look something like a sunflower, and will keep its mirror always facing the sun.

What the people who land on Mars will find there (Carl Sagan, *The Cosmic Connection*, loc. cit. has covered this matter thoroughly) is anyone's guess. I think they will find mosslike plants, a few places where there will be ponds of water during some of the Martian seasons, and little else of interest. The importance of Mars probably derives more from what we can make of it than what it is (see Chapter 3).

As work on the Mars landing goes forward there will be a gradual but decisive change in the nature of the activities of the work force in space. Some of the workers, when their contract expires, will elect not to return to earth but stay to make a career in space. This decision will be irrevocable. After a presently unknown period of time, which will no doubt vary a great deal from one individual to another, and which may be of the order of a year, the changes in the body incidental to living in the absence of gravity will become so great that one can no longer readjust to life on earth. This may seem to contradict what we have already experienced. If a man can live thirty-odd years on earth and then, in a few minutes, adjust to life in space, as several dozen astronauts already have, why can't he do the same thing in reverse? The answer is that adjusting from a life under gravity to a life without gravity is a process of unloading. Without gravity the heart has less to do — it no longer has to pump blood up from the feet to the head — and the bones have less to do too since they will not have to support the weight of our organs. It will also be easier to go from place to place under our own power in space than it is here. Swimming at low speeds is easier than walking. Readjustment from space living to earth living is a process of loading. Our astronauts have done it without much trouble after a month and more in space, but it must have seemed like putting on a one hundred fifty pound pack. Some of us could do that, perhaps, and carry the pack for the rest of our lives, but would it be worthwhile?

Meanwhile the recruitment of younger workers will begin; these will be committed, even before they go out there, to a career in space. This group will contain both sexes. They will need training only in health and survival procedures before they go out; the technical skills they need will be taught after they arrive.

The effort which can be spared from the weather control, space power plant, manufacturing, and Mars landing programs will be devoted to reducing the load on the shuttle service by reducing the dependence of the colony on supplies from earth. The main thrust in this direction will be toward growing plants in space. Within a dozen years of an operational space shuttle these plants will provide all the oxygen and more than half the food consumed out there. The biological and engineering problems involved in doing this are so complex that little more can be done with them in earth based laboratories; the practical system must be worked out by trial and error in orbit.

Perhaps even before the Mars landing an event of extraordinary importance will take place in space. The first child will be born there. Our descendants, a very long time from now, may not have heard of the war in Vietnam, or of Watergate, but they will know of three great events which happened in this century — the placing of the first object in orbit, the first step on the moon, and the birth of this child, just as we now remember the birth of another child two thousand years ago, while forgetting the military and political events which were happening then, and which were so much more important to the people then living than the birth of the child.

Notes

1. I am not concerned at this point with education in the liberal arts. The great thought of the past, with its art and music, is not becoming obsolete, but we should now formally recognize it for what it has always been — a luxury available only to those who can afford it. If we are to change the system of teaching the liberal arts at all we should perhaps better revert to the way it was taught forty or fifty years ago. We should up the tuition at liberal arts prep schools and colleges to double what it actually costs to run them so that a liberal arts education can be available to a very few highly gifted young people who cannot otherwise afford it.

Enrollment in a liberal arts course should be recognized for what it is, a gesture of defiance, an assertion by the student or his parents that he has so much inherited ability, such self discipline, and is so intelligent, that he can give his less endowed contemporaries four years start and still beat them in the competition of life. We shall have to do what we can to prevent enrollment in a liberal arts course being used as a status symbol by parents, though here we shall fail. We all see young lives ruined by trying to force the student to become something he is not and never can be. Though we cannot eliminate this abuse we can perhaps minimize it by making it extremely difficult for anyone to enroll in a liberal arts course. We may, for financial reasons if for no other, have to eliminate most colleges now giving liberal arts degrees.

2. Parents should still be allowed to bet, as they do now, on an unpromising child, but it should be an expensive bet, and the child should be further schooled in private institutions.

I made such a bet and won. I was solemnly informed by the school authorities that my son was not college material. He graduated from Yale.

3. Geometry and arithmetic are in much the same relationship as male and female people, much alike, complementary, but with subtle and precious distinctions. They should be kept at a certain distance until they are properly married in the theory of functions of a complex variable, after which the union can be fruitful.

4. When I find myself in logical difficulties, as I do at this point, I look for a numerical way out. I have found one, and here it is: If we could get the birthrate up to an average of seven children per family and keep it there indefinitely, the proportion of old people in our population would diminish and steady at less than ten percent of people over fifty, even if no one died. This might be bearable.

For those who wish to delve deeper into the numerical possibilities of this sort of thing I note that the e-folding time for increase of population is found by dividing eighty-six by the average number of children per family less two point three. From this the fraction of the total population which is over a certain age is found by taking the negative exponential of this age divided by the e-folding time. This calculation takes no account of deaths; if the usual actuarial tables apply the proportion of older people will be even less. The e-folding distance, in either time or space, is the interval in which a quantity increases or decreases by a factor of $e = 2.71828...$ This weird number, which never comes to an end and the order of whose digits never repeats, is the limit of the sum

$$1 + \frac{1}{1} + \frac{1}{1 \cdot 2} + \frac{1}{1 \cdot 2 \cdot 3} + \frac{1}{1 \cdot 2 \cdot 3 \cdot 4} + \dots \text{ as the number of terms}$$

is indefinitely increased. It is found in many mathematical expressions for many reasons. As an example, if you compound interest continuously, instead of by the day or by the month, the accumulation is found by raising e to the power obtained by multiplying the rate by the time.

5. Even in the United States, one can find, if he looks for them, people who are undernourished, but this is nowhere for lack of the right kind of food in the community. Most often it is due to lack of information as to where food is provided free to those who really need it, or not knowing how to prepare food in an effective way, or of what constitutes a healthy diet. In some areas it may be due to ethnic or religious prejudice, but these are political, not technical problems. What is meant here by starvation is absolute lack of enough proteins of the right type to permit a balanced diet or lack of enough food of any kind to add up to enough calories, these shortages prevailing throughout a whole geographical area.

6. The often quoted statistics which show that college graduates earn more than those who do not go to college are correct but they are often misinterpreted. What these statistics prove is that young people sufficiently intelligent and self-disciplined to become good wage earners are more likely to go to college than those less gifted in these respects.

7. It was at one time thought that the goal of automation was to get us more leisure. If the machines do it for us we might have to work only one day a week. I, and probably many others, would prefer more leisure, but it has been found that most of us, if our job requires less than thirty hours a week, take a second job along with it — the process known as "moonlighting".

8. When your letter comes in, a girl reads it and punches some holes in a card. She cannot deal with your problem herself, because she doesn't have the necessary information. There are only a limited number of places on the card where she can punch, and she has to fit your thoughts into these as best she can. The computer has only the card to go on. From this it decides what information is needed to answer your letter, digs this out of its memory banks, decides which form let-

ter is applicable to your situation, and enters what it thinks is the relevant information in spaces provided in the form letter. Girl and computer agree on one thing — the fact of answering you letter is more important than the specifics of the answer.

One ploy which often works is to visualize the card which the girl will punch and write your letter so that she can't see quite how to do it. The object is to confuse her enough to buck your letter up to her supervisor but not enough to make her throw it in the waste basket. The supervisor does not punch a card. He types your information into the computer from a keyboard and instructs the computer what information is relevant. The computer then displays this information on a screen and the supervisor decides how to solve your problem. Sometimes he even writes a personal helpful letter about it.

9. Detailed Statement for the Record by Dale D. Myers, Associate Administrator for Manned Space Flight, National Aeronautics and Space Administration, to the Committee on Aeronautical and Space Sciences, United States Senate, February 26, 1974.

10. It has been suggested that before going on to Mars we should establish a military base on the moon, or at least take steps to prevent others from doing so. The moon has no value as a military base. It is true that a rocket fired from the moon can hit any point on earth. It is equally true that we and everybody else can see such a missile coming two or three days before it gets here, and launch counter-missiles against it. The missile may have some capability of shifting point of impact shortly before it enters the earth's atmosphere, and we cannot know precisely where it will hit until a few minutes before it does so, but it cannot change the time of impact and, with a day or more's warning of a civil defense emergency we can get almost everyone safely under ground in two or three minutes after we know the point of impact. The important thing is that if anyone fires a missile at us from the moon we will fire back, not at the lunar base but at their country, with missiles which will take half an hour instead of a day or more to arrive. Any missile fired at us from the moon will arrive after the war is over.

11. It is commonly supposed that we were able to go into Japan unopposed because we dropped two atomic bombs on them. No doubt this did hasten the capitulation, but as one who, soon after the events, compared on the ground the damage to Nagasaki from an atomic bomb and Yokohama from conventional bombs, I can testify that Yokohama was the more thoroughly destroyed. Without the atomic bomb the Japanese would probably have offered some resistance to our landings, but they had been so battered during our methodical approach that they had little left to fight with.

9.   The Evolutionary Process

The reader is no doubt acquainted with the main lines of the evolutionary process and with its history; with those who came before Darwin, notably his grandfather, who thought about the same things and almost made the synthesis but not quite; with the voyage of the Beagle and young Darwin's observations, with his doubts about the significance of what he had found, and finally with his unwavering conviction that he understood what had happened. I shall not comment on this history except to point out an apparent tautology. Darwin summarized his ideas as evolution being the survival of the fittest. But the fittest for what? The fittest to survive, obviously. Those who survive are the fittest to survive. One needs to add here, those fittest to survive in the short term, in an environment unchanging over many generations.[1] Up to now every species has had a period of unchanging environment, but that time has passed. Our environment changes decisively and dramatically during our own lifetimes and we need to raise again the question "the fittest for what?" My answer is "the fittest to colonize."

What I hope to accomplish in this chapter is to identify the space program as an essential part of the evolutionary process, partaking of the inexorability of evolution, to develop the concept of *force majeur* in this context, and to show how acceptance of the identification and the force clarifies our thinking about ourselves and our destiny.

Our evolution so far, from one celled sulfur breathing plants to ourselves, has been primarily a chemical process which appeared to have come to a dead end about ten thou-

127

sand years ago. Chemical evolution could and did bring us to the point where we could dominate the surface of the earth and to some extent change our environment here, but it could not get us off the surface of the earth. This could be accomplished only by developing an electromechanical capability. That the apparent dead end could be passed illustrates the wonderful subtlety of the evolutionary process. To get from chemistry to electromechanics we needed the new capability of imagination which animals have only to a limited extent if at all. The new capability did not develop directly. The starting point was curiosity, which the animals have, and this led to storytelling and pictorial art. These in turn led to written languages, and these to classification of knowledge. The existence of bodies of knowledge led to generalization, and this led to philosophy and more sophisticated art, literature, and music. The required new quality of imagination was thus fully developed, and then applied. The application produced science and technology, and these produced the required electromechanical capability.

The history of evolution is a history of colonization. A species perfected itself, in the short term, by selective breeding and survival of the fittest; it improved its chances of survival, in the long term, by migration from the place where it was to a place where it could do better.[2] The one celled plants deep in the mud of the sea bed found they could do better where they could get some sunlight and moved up to the top of the mud in the shallows where sunlight could get through to them. The more vigorous oxygen producing plants which did this found that they could do better still by moving up into the water itself, and did. Some of these developed a limited capability of swimming and began to feed on their neighbors. When they could do this they no longer needed sunlight and went back and recolonized the mud.[3] From these came our first animals, with powers of locomotion. The animals soon found their environment overcrowded, with not enough food for each to eat as much as he could have used, and so the animals began to colonize the hostile environment of the tidal flats and beaches, and did well there, in spite of trouble from waves and tides, feeding on things they could not reach before, and with some escape from overcrowding. Soon the flats and beaches were overcrowded too and the animals began to colonize the shore line above the high tide level. They were able to do this, and to get into less crowded pastures, but for a long time they had to go back to the water to reproduce, as frogs and toads must to this day. Finally this limitation was overcome by development of the tough-skinned egg, and there were animals who could live independently of bodies of open water. Then the land became overcrowded and some species took to the air. At that time life was practically limited to the sea and the low lands of the tropics; life in the mountains and in the harsh environment of the temperate and arctic regions was beyond the capability of the species then existing. Some of these gradually developed hair and feathers to keep them warm, and a temperature regulating system which permitted them to function efficiently in uncomfortably hot and cold environments. The last place on earth to be colonized effectively was the area around the South Pole. Only man could do that.

I find the history of recolonization of special interest. The mammals, after immense struggles as fish, amphibians, and lizards, managed to free themselves of the restraints of an aqueous environment and warm environments so that they could live anywhere on the land, but some of them returned to the sea to become whales and manatees (a distant cousin of the elephant). The sea otter[4] is making the transition back before our eyes; he is a complete and capable land animal but can spend his entire life in the sea. No catlike creature has gone back to live in the water.[5]

We, too, could go back to a completely marine life with the aid of some gadgetry which is now available. Sea water, at the pressures prevailing at depths of fifteen hundred feet or more, will dissolve enough oxygen to be breathed in the lungs instead of air. Experimental rats and dogs have already done this. You drown them in water saturated with oxygen at these pressures and they don't die, but just go swimming around in it, breathing water instead of air. No doubt we could do it too; after all our lungs have evolved from gills and were gills at one stage of fetal development. The catch is that it seems to be a one way trip; no experimental animal has yet made it back to breathing air. Also, once committed, one could never come up to depths less than fifteen hundred feet and the only light he would see again would be artificial or the faint glow of luminous fish. We would need electrically heated clothes to keep us warm, a supply of oxygen to saturate the water we breathed, a filter to keep our lungs from filling up with mud, and fish-eye contact lenses to adjust our vision to the refraction of the water (a skin-diver's mask could not be used because there would be no air to fill it). Our homes would be aquaria where the water would be kept warm enough so that we could take off our clothes and well enough filtered so that we could breathe it.

The mammals who have returned to the sea have enhanced capabilities; they make better use of muscles and blood and have better reproductive processes than their ancestors, the fish, who never left the sea, have been able to evolve. Perhaps the earth will someday be recolonized by descendants vastly more capable than ourselves who may bring about the miracles of culture our present intellectuals dream about.

Colonization has never meant that everybody goes. Only a tiny fraction, consisting of hardy souls, of the total population emigrates. The bulk of the population remains behind much as it was.[6] Horseshoe crabs have never migrated, but they seem to be doing well; however, most species left behind finally die out. For every species existing today more than a hundred have vanished, perhaps because they ceased to be colonists.

I recapitulate here some aspects of the evolutionary process which seem to be relevant to the space program.

Evolution is chaos. It is the antithesis of a plan. Everything which is genetically possible happens, not once but many times. Almost everything that happens is absurd, grotesque. But a few things happen which further the mission and these are the things which happen ad infinitum. It is these very few things which give the process not rationality but purpose.

Evolution is waste and prodigality. A cod in her lifetime

produces more than a million eggs of which at most two mature to individuals as large as she. Billions of lizards had to live and die aimlessly before a few of them developed hair and a few others feathers. More than a hundred pounds of food must be produced by hardworking ocean plants to put one pound of fish on our tables. Thousands of potentially great minds[7] never develop while two or three of them become known to us as a Shakespeare or an Einstein.

Traditionally waste is anathema to us, partly because our ancestors a few generations back had such a difficult time scraping together the few possessions and little food which they had, and partly because science is built on the concept of efficiency. We engineers are so thoroughly indoctrinated with the imperative of efficiency that I found it difficult to bring myself to design electric motors which ran hotter and burned more power than they needed to just because such inefficient motors better served the purpose for which they were intended. We are conditioned from birth to hate waste in a world created by waste.

Evolution is tragedy. Birds lose almost all their chicks. Squirrels die of freezing. Rabbits starve. All animals die in conditions of acute distress. The swan tries to sleep in sleet and rain. There is never enough for an animal or a bird to eat. Many people are often cold and hungry too. Some live in terror of military situations. Some mourn lost children. Some are disappointed in love. Affluence spares no one. A man is distressed because he has no worthwhile friends or because he has not the prestige and power he thinks his due. Even those few like myself who have had immensely satisfactory and pleasant lives at every period of age have to give up pleasurable activities one by one because of waning physical powers and have to go just as they are beginning to understand life and how to enjoy it. Tragedy is the central theme of every life, and if there is any difference at all in the quality and intensity of individual tragedy it is not large.

Evolution is failure. For every species of animal on earth today a hundred have become extinct because they could not compete successfully with other animals. Some — perhaps all

— species except ourselves which have not died out will evolve into dead end situations in which any further modification would be unfavorable competitively. The whales, after a circuitous trip to land and back, have reached their own watery dead end. These and many other species have failed and are just waiting for some new creature or event to finish them off. We may be in the same category — perhaps not.

The paradox is that only such a chaotic process as evolution — wasteful, tragic, and hopeless as it seems — has any chance of succeeding in a grand design. The alternative is a planned approach of divine or human origin. Every design, planned or not, is subject to chance deviations or departures whose nature and timing cannot be known in advance. Planned agriculture, for instance, is subject to unpredictable deviations due to weather. In a plan such a deviation inevitably leads to a departure from the original objectives — a flood diverts effort away from producing food in the direction of controlling floods. Such departures compound each other and soon destroy utterly any progress toward the original objective. But a chance deviation from chaos can lead only back to chaos. Chaos is stable; when, in evolution, or the free expansion of a gas, chance deviations occur the original objective — or inevitable outcome, if you will — is not diluted by these chance fluctuations. Thus the objective is reached rather than gradually abandoned.

This is why democratic regimes are so much more productive of both goods and ideas than authoritarian regimes. An authoritarian regime plans, and in patching up its plans to cope with unexpected factual situations departs ever more from the objectives of these plans. A democracy cannot plan at all except in an emergency so desperate that everyone agrees what must be done, and the plan collapses as soon as the emergency is passed. If a politician speaks of objectives in terms more specific than the vaguest platitudes his plans are immediately chopped to pieces by an opposition. By keeping close to chaos but never quite achieving it we gain stability and time in which great projects come to fruition and great thoughts are elaborated for posterity.

132

It is my contention that it is only in the context of evolution that human life makes any sense at all. Any other approach to rationality leads to frustration. Religions and philosophy don't stand up to close scrutiny in the light of present knowledge. All of them promise some reward for being different from what we are and these hypothetical advantages become less credible daily. Morality and value concepts have been totally lacking in stability. We shift from the far right to the far left, from tyranny to excessive permissiveness, from prudery to license, from economy to profligacy, at unpredictable times and for obscure reasons. Every possible combination of morals and values has, at some time or other, been convincingly urged by some great man and applied for a time by a considerable number of people. Philosophic theory has shifted in time and place, sometimes by reason, sometimes by war, sometimes by the discovery of new lands, and sometimes by the discovery of new scientific knowledge. There has been no pattern to such change; history is chaotic. The only aspect of history which has remained unchanged from the earliest records to the present time has been an increase — sometimes fast, sometimes slow, but always positive — in technological capabilities. This has been the path through chaos — the only thread to which a rational man could cling.

All through history people have looked for reasons for the things which happen and the things which they do, but no reasons which will stand up to scrutiny have ever been found. Now the experiment is complete, the results are in, and we know that there are not, and cannot be, reasons of this kind. The concept of reason must be replaced by a concept of destiny.[8] We are what we are and we do what we do because we are part of an inexorable process. This is what is meant by *force majeur.*

The idea of *force majeur* as applied to ourselves is difficult to accept, and we probably could not accept it but for the fact that we have another *force majeur* as a precedent. Prior to 1821 science was a collection of separate disciplines — mechanics, chemistry, optics, electricity, acoustics, to name a few — related only by the fact that in each of them resort to experiment was the ultimate authority. In 1821 Nicolas Carnot published his *Réflexions sur la puissance motrice du feu* and scientists started thinking along different lines. The result was the identification of the *force majeur* now known as the laws of thermodynamics and the transformation of science from a heterogeneous collection of facts into a unified body of knowledge. The details of how this happened are of interest and relevant to the present inquiry.

Ever since the appearance of the steam engine people have been trying to devise machinery which would produce more power than it consumed. They tried to use a steam engine to drive an electric generator and to use some of this power to electrolyse water, producing hydrogen which could be burned in the boiler to run the steam engine, and use the rest of the power for something else. There have been many ingenious proposals of this kind but none of them have worked. There has been a great deal of fraud and quackery associated with these efforts, analogous to the fraud and quackery which have always been associated with philosophy and religion. But failure to produce one of these perpetual motion machines proves nothing in itself. There is always the chance that a smarter and better financed engineer will come along and do it.[9]

The scientist Rudolf Clausius took a bold step when he turned the problem of perpetual motion inside out — inverted it, as mathematicians say. "Let us take this thing (the impossibility of perpetual motion) which we can neither prove nor disprove experimentally, as an axiom" he said "and see what we can deduce from it." This was the identification of the impossibility of perpetual motion as a *force majeur*. It led Clausius to the concept of entropy, a property of matter utterly unlike any previously studied, in that it could not be directly sensed or measured by any instrument, and did not remain constant in total amount, as mass, momentum, energy, and electric charge did. According to Clausius, the energy of the world remained constant while the entropy of the world drove toward a maximum.[10] The new concept was used by Lord Kelvin to explore the definition of temperature,

134

and by Professor Willard Gibbs at Yale, Doctors Nernst and Haber in Berlin, and Professor Theodore William Richards at Harvard, to predict chemical equilibria on the basis of purely thermal measurements. It provided a missing facet of geometrical optics by showing why, as had long been known, it was impossible for any combination of lenses to increase the brightness of an image. Boltzman, in a brilliant paper, showed how to use these ideas to calculate the pressure exerted by a ray of light. His result was in good agreement with experiment. Finally, in 1929, Leo Szilard made the identification of entropy with information, and gave the communications industry its most powerful tool.[11]

What thermodynamics did to unify science, the recognition of the inexorability of evolution and our part in it, may be able to do to present day economics, sociology, politics, and technology from a unified point of view. This is for the individual reader to judge. If acceptance of this *force majeur* clarifies his thinking about world problems, fine. If this point of view confuses more than it clarifies it is not for him. I give below several examples of how it has clarified my thinking.

Take the question of policy toward the environment. If we accept the proposed *force majeur* we see at once that our destiny is not to preserve the environment but to change it. We have been changing it since we were one celled plants and we will go right on changing it, and it us, until neither we nor the Milky Way are recognizable for what we once were.[12] When we see this we recognize the present conservation movement as a cult with dogmatic imperatives. We have had a lot of experience dealing with cults, and we know how to respect the beliefs and taboos of the individual members of the cult without allowing ourselves to be dominated by them. Finally, we recognize the problem of the environment as primarily one of aesthetics and secondarily one of economics but not in any way a problem of health or survival. Conviction that the destiny of man is to change the environment in no way detracts, and perhaps actually enhances our appreciation of natural beauty and our desire to preserve it so far as we can within our ancestral imperatives. It becomes a question of determining, on aesthetic grounds what we want, on political grounds how we can get it, and on economic grounds how much it will cost and how much we, as individuals, are willing to pay for it.[13]

Now for an application to the drug abuse problem. Evolution operates, in the short term, by survival of the fittest,[14] and if it fails in the short term it cannot succeed at all. Therefore, by *force majeur*, if mechanisms which have in the past biased the social and economic systems in favor of the fit disappear, then inevitably new mechanisms will appear, or old mechanisms will be expanded, to restore the bias. I think drug abuse is one of those old mechanisms we have always had which has expanded to fill the gap left when we did away with predators and diseases. If we wish, as I do, to reduce drug abuse we must provide some other mechanism to insure the survival of the fittest. This might, for instance, be compulsory sterilization of everyone twice convicted of a felony. This does not imply that we should refrain from presently employed means to reduce drug abuse — these have their place — but suggests that unless we simultaneously adopt a more fundamental approach to the problem these will fail — as they are failing now.

Finally, in a lighter vein, consider the problem of services and wages. An electrician, with little investment in schooling, wants to be paid as much as a college professor. Ridiculous? Not at all, from the evolutionary point of view. Our scholars and contemplative people have done their thing — they have gotten us around the evolutionary dead end from chemical to electromechanical capability — and we do not need them any more. To go further we now need a race of doers, not thinkers. We have too many thinkers and we need to pinch some of them out of the economic system. From this point of view the electrician, greedy, incompetent rascal that he is, plays a role in the grand scheme of things.

Notes

1. "Many generations" may not seem to be "short

136

term". For man many generations might mean a thousand years, but man has existed as a species for fifty thousand years. Thus, with respect to man, a thousand years may be taken as short term and anything over fifty thousand years as long term.

2. The horseshoe crab, a distant relative of the spiders, never did this, and remained in the same form and place for more than five hundred million years. It is at a dead end and may stay there for a very long time yet.

3. The fungi are remnants of this population.

4. The seal and the sea lion are on their way back to a completely marine life, but they still have to come ashore to reproduce. The sea lion has been so recently a doglike creature that he still barks.

5. The sea leopard has not evolved from a catlike creature but is just a large and vicious seal.

6. The migration of the Goths may have been an exception.

7. One often hears, usually in connection with education, what a dreadful thing it is to waste a mind, and how much we now need great thinkers. It is in fact quite impossible to waste a mind and if we produced a lot of great thinkers we wouldn't know what to do with them. Our culture can absorb at most three or four such great thinkers per century, and if we produced more than that we would just have a lot of frustrated people. The fact that millions of potentially great minds never develop is part of the waste inherent in the evolutionary process. Every worker bee was potentially a queen, and in one year, in one hive, perhaps ten actually develop into queens, and then have to kill each other off until only one is left.

8. This is not a new idea. It has been advanced by the philosopher Spengler and others.

9. There are devices which produce more power than they consume — for a while. The battery in your car is able, if discharged slowly, to absorb heat from the atmosphere while delivering power more than equivalent to the heat absorbed, but it cannot do this indefinitely. Strictly speaking, the rule is that no machine can deliver indefinitely more power than it

absorbs from outside sources. To complete the statement of the scientific *force majeur* we must add that no machine whose only source of outside power is heat can produce more power than the net heat absorbed, and that any such machine must absorb heat at a high temperature and reject part of it at a lower temperature.

This *force majeur* is sometimes stated as "perpetual motion is impossible." This is probably true as applied to devices big enough for us to see, but there is reason to believe that perpetual motion does occur inside atoms, and perhaps in larger aggregates of matter at very low temperatures.

10. "Die Energie der Welt ist konstant; die Entropie der Welt strebt einem Maximum zu."

11. I learned personally from Nernst, Haber, and Szilard, and had, at the time of his death, been assigned to do research under Professor Richards.

12. Modern technology, with its parking lots and high rise buildings, has not yet affected the environment as much as the activities of plants, animals and primitive man. One celled plants built the White Cliffs of Dover, and coral animals and plants have changed shorelines throughout the tropics. Deforestation by primitive men in the Fertile Crescent and in China have done more to change the face of the earth than anything since.

13. For application of these ideas to the stripmining problem see appendix 3.

14. Survival of the fittest does not mean that all the fit survive and all the unfit perish. It means only that fit individuals have a slightly better chance of surviving than unfit individuals. In the operation of any actual selection mechanism some fit individuals will be struck down and some unfit individuals will survive to beget more of them.

10.   The Human Material

It is time now to take a hard look at ourselves to see whether we are adequate for the tasks ahead, and if we are not, to see what we can do about it. For if we are to survive as a species and really move out to the stars, it is we who must face the problems and carry the burden. Machines will help, as we have seen in Chapter 7, but they cannot do it without us. In looking at ourselves we can no longer duck the two fundamental questions of philosophy and religion: "Who are we?" and "What are we doing here?"[1]

Searching for these answers we shall look first in the past. We do this because we of the West have elaborated our heritage from the ancient Greeks into a philosophical system which we term rational, and we think things out according to the program of that system, as a matter of habit. We know that there are other philosophical systems, differing from ours, and used and trusted by large numbers of people, notably the Hindu, Buddhist, and Marx-Leninist philosophies, and that these systems, when applied to decision making, give answers that differ from ours; but we reject these other systems because ours works when applied to the physical sciences and the others don't. Our system requires us to look to the past before speculating on the future. This we now do. If we reject the "lessons" of the past as applied to our problem we owe it to ourselves to know what we are rejecting.

The present religious systems are distillations of the wisdom of the past and we look first to them for answers to the fundamental questions — to three very old religions, two of intermediate age, and two quite modern. To the Jewish, Chris-

139

tian, and Moslem philosophers we are the image of God and our purpose here is to worship and glorify Him. To the Hindu we are an evil spawn, existing only by the munificence of the Brahmins, and our purpose is to die off as quickly and unobtrusively as possible. To the Buddhist we are pathetic doomed things, hovering between the human and animal worlds, and our purpose is really to die instead of just going through the motions of dying. To the Marx-Leninist we are indistinguishable as individuals but part of a mass — the good mass — and our purpose here is to defeat and govern the people who constitute the other mass — the bad mass. To the Maoist we are prudent farmers, dedicating our lives to uprooting the weeds of bureaucracy, secure in the knowledge that no matter how many such weeds we uproot we shall end up with more bureaucracy than we had to start with.

These philosophies all assume that the nature of a man can be changed by indoctrination, or religious exercises, or both. The fundamentalist Christians and the Marx-Leninists have gone to extremes by way of indoctrination and the Zen Buddhists by way of religious drills. These procedures have not worked, and human nature has not changed since the earliest recorded history. To see this one has only to read the Bible and the Vedas;[2] people then did exactly the same things in exactly the same ways for exactly the same reasons they do now. Put Jonah back in the whale and cast him up on Jones Beach and he would never know he had left Nineveh. There is some evidence from the evolutionary process (Chapter 9) that the nature of the human species is changing, and will continue to change if we survive, but only very slowly — perhaps a hundred thousand years to produce a readily discernible change — so that it is not astonishing that we find no difference in the scant four thousand years of recorded history, and cannot plan on it for the few hundred years ahead.

I propose that, for the purpose of thinking about the space program, we discard historic concepts of what we are and why we are here and substitute empiricism based on observation of the present. This is what we do in science and engineering. We do not think of glass as bad because it shat-

140

ters and good because we can see through it, or steel as bad because it is opaque and good because it does not shatter; rather we study and measure the properties of glass and steel to see how we can use them. The concepts of good and bad are wholly absent from the physical sciences. So I propose that we study the properties of the human material, not by reading learned treatises about it but by looking around at our friends and associates to see what they actually do and what they actually want. As we do this we find we are greedy as newly hatched robins, and the more we have of anything the more we want of it and the harder we will fight to get it. We are aggressive, usually to get territory or things away from others but sometimes out of pure exuberance. We are curious about small deviations from the norm, such as collisions between automobiles, but shy away, to the point of revulsion, from thinking about large deviations such as nuclear war. We are acute at detecting small untruths but suckers for large ones, as Hitler pointed out. We are cruel and kind, each capriciously; decent and indecent, each to extremes.[3] It was us the prophet Isaiah had in mind when he said "And you, happy city, turbulent city, your dead are not dead of the sword, or dead in battle!" We are scum by any conventional standard, but scum which has come from the bottom of the sea, up over the beaches, through caves and cities to the moon, and which is capable of going much further still. We are a lusty, bawdy, unruly lot, and I hope we shall always remain so.

It has been suggested that we change ourselves into something different, and presumably better, by genetic engineering. This could take either of two lines, selective breeding or laboratory modification of genetic material, and probably both together. I think that this could be done, probably within a few decades, but for the following reasons I do not think that we will wish to do it.

If we are to be changed by genetic engineering into something better we shall have to arrive at some consensus as to what we want to become, and I do not think that this is possible. Our individual preferences, and those of our cults, ethnic groups, and economic blocs, are too diverse to permit this. If I had to decide the question I should first rule out any superbeing who was too smart. I would not want something out of a test tube explaining in detail where my thinking was wrong and perhaps deciding, in an offhand way, to do away with me altogether. I realize that there are some who look to superbeings for greater music, literature, and art, but I can wait for that, sticking with Bach, Cervantes, and Velasquez for a little longer. A really good superchemist who could deal with the problem of aging, or a superphysicist who could figure out a way to avoid wrecking interstellar ships on gas clouds, might be acceptable if he is programmed in such a way that he cannot possibly concern himself with social problems.[4]

In the distant future, when we are firmly established in space, we may want to create some specialized types who can work in environments too tough for us. Sulfur breathers, perhaps (after all, the Devil is a sulfur breather) for planets where it would be difficult to supply oxygen, or hibernating types who can simply go to sleep when it gets too cold and wake up in fine shape when it gets warm again. These specialists should be smart enough to comprehend operating and instruction manuals but no more. We may need their services but we would rather be spared their philosophy.[5]

The deeper reason for rejecting the genetic engineering approach is that we are, just as we stand, ideally suited to colonize distant planets. It is hard to see how genetic engineering could make us any better for that. As shown in Chapter 9, the thrust of the evolutionary process has been to produce successful colonists, and we are at the point of that thrust. In engineering we know that the device perfected by trial and error is always better than the device which springs complete from the engineering mind.[6]

There is a way, less drastic than genetic engineering, by which we can improve our performance. This is to get rid, once and for all, of the concepts of guilt[7] and shame, and, so far as possible, to get rid of hypocrisy. I am told that guilt feelings are the most destructive of all human emotions. I cannot

142

vouch for this because it has been many years since I have experienced guilt feelings of any kind. We can admit, unrepentant and unashamed, to what we actually are and dismiss as trivial and irrelevant those nagging thoughts that we should, as we have been told all our lives, be something different.

Hypocrisy can be a crushing burden. Participating in the invasion of Japan I was astonished to find that the Japanese bore no grudge and were simply relieved that the war was over. I was even more astonished to hear them say that while they were glad that the bombing had stopped they were even happier that they no longer had to pretend that they were supermen and even divine. Intuitively they had known all along that they were nothing of the kind, and now that this had been publicly acknowledged they could go happily back to being Japanese.

The extent to which we go on with the space program will be determined by what we really want — not what we think we want or what we say we want but what our actions show that we want. As I interpret what I have seen people actually do, we want primarily the right to decide small matters for ourselves but not to bear the responsibility for larger decisions. We want to choose whether or not to have onion rings with hamburger, whether to wear broad or narrow lapels, what kind of car to drive, and the more vigorous among us may even want to decide where to live. Larger decisions, such as monetary policy, foreign relations, and how much money to spend each year on the space program, we want made for us by an authority so remote and inaccessible that its decisions cannot be challenged. We feel the need to respect authority but not to like it; to comply with the dictates of authority but not necessarily to approve of them.[8]

We feel the need to be identified with something larger than ourselves and to be enfolded in it. We need a goal in life which we can understand clearly and which we can actually reach, even though we may not like it; not the fuzzy goal which we have been brought up to believe we ought to seek,

143

but which we know intuitively is fundamentally indefinable and forever beyond our reach, except perhaps on some distant planet.

Such authority, goal, involvement, and freedom from having to make troubling decisions, the space program can provide, but there is a deeper reason why we may choose to go on with it. Many of us, as we cope with the perplexities of our daily lives, think back to the good old days on the frontier, when men were strong, women chaste, and anyone with a well muscled back was as good as anyone else. Our predilection for Western films attests to this; even the Italians and the Japanese, who have no comparable tradition of frontier days, make their own Westerns. Kindness and generosity, those properties of the human spirit which we prize most, were more evident on the frontier than they are now. We cannot have new frontiers[9] on earth, but we can have them in infinite succession on distant planets, and we may find there the warmth, love, and human kindness for everyone which we so ardently desire here and cannot have.

Notes

1. It has been suggested that we look inside ourselves for answers to these questions. I have looked inside. There is nothing there.

2. The Buddha himself placed his faith in indoctrination and rejected religious exercises but his disciples reinstated these even before he was dead. The Hindu philosophy, although essentially nihilist, abounds in religious exercises, including Yoga and Transcendental Meditation, now widely followed in the United States. All religions are layered structures. Those in the top layer of the hierarchy believe nothing and despair of the prospects of the human race. Those in the bottom layers understand nothing and can move upward only by means of religious exercises blindly followed.

3. We can add to this list of proved attributes. The lower a man is in the hierarchy the more he oppresses those below him. No one hates and despises the poor like the poor. We are

all gamblers, and we like to play long shots rather than close ones. Because we were so recently hunters we are fascinated by violence and are thrilled by watching events in which someone may be hurt.

4. We shall soon be forced to selective breeding, no matter how little or how much money we put into the space program, but for economic rather than genetic reasons. As technology comes more and more to dominate our daily lives, increasing numbers of individuals, for lack of an adequate combination of intelligence and self discipline, are unable to cope, and become a dead weight on society. Except for selective breeding, the only way we can accommodate these people would be to go back to subsistence farming, and we are not going to do that here, though we may do it later on distant planets. We love these improvident people dearly but there are just too many of them now and we can't afford them any more. Perhaps we need a constitutional amendment defining the status of the improvident, making them wards of the State and relieving them of the responsibility of voting, having children, or organizing into unions, and taking them out from under the shelter of the minimum wage laws.

5. Genetic engineers could begin by reprogramming some present insects: moth larvae, for instance, to replace our too loquacious barbers. Or perhaps silk worms and spiders (technically not insects) to weave fabrics or assemble microelectronic equipment. It is silly to use our clumsy fingers to deal with small things like thread and wire. Or perhaps stingless wasps to make paper out of wood.

6. This may seem to be at variance with recent engineering history but the contradiction is only apparent. The necessities of World War II created the magnetron, without which there could have been no effective radar, and the rocket and the atomic pile. There was no significant previous experience with any of these, but they were just the brilliant exceptions which prove the rule. Each of these performed well just as it left the drawing board. The television set and the computer, which play such important parts in our daily lives, came along just before World War II and did not do so well.

145

The consumer is still perfecting them, painfully, by trial and error, as all of us realize when we tackle our monthly bills, and both have a long way to go before they attain the reliability of the consumer tested automobile.

7. If there is to be no more guilt what can we do about restraining crime? I think our penal system would benefit from the renunciation of all idea of guilt. "We don't think you are any better or any worse than we are" we can say to the criminal "but there are many more of us than there are of you. We don't like what you are doing, and we are prepared to take any necessary steps, however severe, to prevent you from doing it."

8. Our fury over the Watergate affair may have derived, in part, from our indignation at having been made to admit publicly what we had known all along, that those who shape our destiny are not supermen but just clods like ourselves.

9. It has been proposed that we have a new frontier beneath the surface of the sea. I think this is within our technical capabilities but I have been down there, at least a little way, and was repelled by the conditions I found — cold, darkness and mud. I would rather take my chances out in the sunlight breathing air somewhat like I am used to. No doubt we shall have submerged mining and agricultural communities, but supportive to our economy on the surface, not part of a grand design.

11. Recapitulation

Here is what I believe has been demonstrated in the foregoing chapters:

1. Even given the best intelligence and good will and the devoted effort of the vast majority of people, the earth cannot remain inhabitable for much more than two hundred years, and probably for less.

2. As people, and as a species, we are not the stereotypes which priests, saints, and philosophers have fed us for the past six thousand years, and which we have all absorbed from birth. To determine what we can and will do, our history before one million B.C. tells us more than our experience since. What we will do derives mostly from the urges of our very remote ancestors.

3. Whatever it is possible for living things, including us, to do, we will do. Wherever it is possible for us to go, we will go. Wherever it is physically possible for human beings to live, we will live.

4. It is possible for people to live and prosper in colonies in orbit about the earth and on other bodies of the solar system, depending on the earth or other planets or moons only for air, sea water, limestone, granite, and phosphate rock. There is room out there for a population more than a thousand times that of the earth.

5. It is possible to build and fly interstellar ships capable of making one way voyages of several hundred years or more to the neighborhoods of the nearer stars.

6. There are, in the Milky Way, about twenty million planets on which the physical conditions are right for life to

have developed during the lifetime of the parent star (about ten billion years). There are several times that many on which physical conditions are not good enough for life to develop, but which could be colonized once we have established a base in their system.

7.   It is possible to land on distant planets from interstellar ships with sufficient people and supplies to found viable colonies, and to build upon these planets populations and industry sufficient to launch their own interstellar ships within five hundred years of the initial landing. This permits leapfrogging from one star to the next until the entire Milky Way has been colonized. This will take about a million years.

12.   Envoi

So now we know who we are and what we are here to do. We are not really people, as our philosophers keep trying to get us to believe we are, but just smart clams on the march, muddling through in a wasteful fashion, preposterous and absurd, vicious and lovable, cruel and kind, and above all free. We are not wracked by the guilt feelings which devastate people; we contain within ourselves the personal tragedy which is the central theme of every life without becoming unduly concerned with the tragedy of other lives. We are free to enjoy the good things which life offers to the lucky few, secure in the knowledge that selfish, greedy and ineffectual though we are, we are still playing our part, doing what we have to do, and moving out in the direction we must go.

Appendix 1.

Design of Orbiting Space Dwellings

The object of this design is living quarters and recreational facilities for a colony of one thousand people in orbit about the earth, and also manufacturing facilities for processing raw materials sent up from earth and making products for sale on earth, as well as parts of new orbital dwellings and interstellar ships. The colony has to be self sufficient, within about one percent of its weight per hundred years, recycling materials except as those are exported as finished products.

The pressure hull is spherical and 260 feet in diameter. Joining the pressure hull at its equator (the polar axis points toward the sun) is a parabolic reflector which makes an angle of 45° with the pressure hull where it joins it, and which extends out to a radius of 184 feet (with respect to the polar axis), where the sun strikes it at an angle of 35°16′ to the tangent plane. The effect of the parabolic mirror is to double the light which comes in through windows in the hemisphere of the pressure hull facing the sun, and to bring this light through the windows at more nearly normal incidence than it would be if there were no reflector. The reflector is subject only to the pressure of sunlight and relatively small forces due to accelerations as the station rotates to keep its polar axis pointing toward the sun, and can be made of extremely thin metal. The total weight of the parabolic mirror is estimated at 5,000 pounds.

Extending backward from the equator of the sphere is a radiating surface in the form of a circular cylinder 940 feet long. Its outer surface is blackened the better to radiate heat. In thermal contact with this surface is a network of aluminum tubes through which n-butane is circulated to carry heat from inside the pressure hull and get rid of it in the radiating surface. N-butane is chosen as the heat transfer medium because of its low freezing point and low viscosity at the working temperature of about −40°F. The weight of the radiating cylinder is estimated at 100,000 pounds.

The hemisphere of the pressure hull which faces the sun consists of a mosaic of hexagonal glass panes in a matrix of aluminum framing. The panes are small—about an inch on an edge—so that if one blows out it can be replaced before too much oxygen is lost—and about twice as thick at the edges as at the center. These panes are dimpled inward because glass is stronger in compression than in tension. The windows contribute nothing to the strength of the pressure hull; the entire stress in this hemisphere must be taken by the aluminum frames between the windows. This framing constitutes about 10 percent of the surface area of this hemisphere.

The weight of the pressure hull is computed on the basis of a working tensile stress of 40,000 pounds per square inch (this is not the figure you would get by pulling bars apart on a laboratory test machine, but what you would get if you pulled sheets of the metal apart while maintaining a lateral tensile stress always equal to the longitudinal tensile stress). With three pounds per square inch internal pressure the thickness of the hull comes out .059 inches, and the weight of the hull, not including windows, 175,000 pounds. The windows add another 90,000 pounds and internal bulkheads 20,000 pounds more.

The contained oxygen weighs 140,000 pounds (computed for 85°F and 70 percent relative humidity). This is more than the community breathes in a year. Machinery, including electric generating equipment, flywheels[1] for keeping the axis pointing at the sun, pumping, and ventilation, adds another 250,000 pounds. One hundred eighty thousand pounds of water are needed, with 3,000 pounds of salt.

To produce and support the plants in the station will require, besides most of the water, some ammonia, most of the sulfur from the sea water, and some of the potassium from the granite brought up. About 500 pounds of phosphorus will be needed, and also about 30,000 pounds of carbon. These will have to be brought up by shuttle.

At the start of the program it will be cheaper to buy and bring up the aluminum, magnesium, and iron required in metallic forms, but high-grade ores of these will soon be exhausted and they will have to be made from granite brought up and processed in space. A typical granite might have the composition aluminum 9.1 percent, magnesium 1.6 percent, iron 2.4 percent, sodium .5 percent, and potassium .8 percent; the balance, calcium, silicon, and oxygen. All of the iron and magnesium, and some of the aluminum, will go into the machinery of the station and the balance of the aluminum will go into the pressure hull, structural members, parabolic reflector, and radiating surface. From the residue of the granite will come all the glass, all the oxygen, and all the potassium needed. A minute fraction of the silicon will go into electronic parts.

Also needed to build an additional station will be some gold, silver, and platinum (perhaps a hundred pounds in all) for making electrical contacts and connections, some nickel, cobalt, and molybdenum to be used in magnetic structures and computer memories, some titanium and tungsten for use in metal parts which have to operate at high temperatures and in cutting edges for machine tools, and some gallium, arsenic, selenium, tellurium, and rare earth metals for use in electronic devices. These are not present in sufficient quantity in granite and will have to be brought up from earth.

Most of the electric wire needed in the station will be made of aluminum but in some applications copper will be needed for this purpose. Small amounts of copper will also be needed as a catalyst in the air recycling system. It is estimated that a total of 5,000 pounds of copper will be needed.

It is now time to consider the energy balance of the station. The total amount of solar energy falling on the hemisphere toward the sun and the parabolic mirror will be close to 13,784 kilowatts (corresponding to a solar constant of 2.0

153

calories per square centimeter per minute). Of this about 10 percent is lost by reflection away from the metallic structure framing the windows, about 3 percent to reflection from the windows themselves, and about 3 percent to imperfect reflection by the mirror, leaving 11,579 kilowatts either entering the greenhouse or absorbed in the glass of the windows. The heat absorbed in the glass is counted as heat absorbed by the station because only a negligible fraction of it is re-radiated. The composition of a glass which might be used is silicon dioxide 71.7 percent, calcium oxide 11.9 percent, sodium oxide 6.6 percent, and potassium oxide 9.8 percent, compatible with the composition of the residue of the granite after aluminum, magnesium, iron, and some potassium and some oxygen have been abstracted from it. This is a low-melting, easily fabricated glass which passes almost all the light in the photosynthetically useful range from green through deep red. We are glad to have some of the ultraviolet light from the sun and some in the far infrared absorbed by this glass as these are useless in photosynthesis. Of the total light entering the greenhouse about 3,460 kilowatts is in the photosynthetically useful range. Assuming a photosynthetic efficiency of 20 percent based on this light (photosynthetic efficiency is defined as the heat of combustion of the solid products of photosynthesis divided by the total energy in the photosynthetically usable light absorbed), this gives 692 kilowatts worth of photosynthetic products, or 692 watts per person in the station. The maximum food intake per day by a healthy person in the United States is 4,000 kilocalories per day or 194 watts. Under the conditions prevailing in this station (85°F. and 70 percent relative humidity) with no one doing any hard physical work, the food requirements might be about half this, or 97 watts, because so little body heat would be lost to the environment. This leaves a comfortable margin of photosynthetic energy to grow some fruit, certainly including bananas and papaya, which are less efficient, photosynthetically, than some other plants, but which will be needed to give variety to a vegetable diet. There is also enough in this energy budget so that we can grow some flowers to brighten our lives, and some cotton to clothe us. The rest of the available energy will be consumed in producing the inedible portions of the plants, some of which can be converted into sugar (as was done in Germany shortly after World War I) and into chemicals. Inevitably there will be some useless residue from these plants which will be burned to recover the contained carbon and water.[2]

All of the heat absorbed in the glass windows or getting into the greenhouse must be re-radiated from the black surface in the rear of the station. If the earth were not nearby this would run about −100°F. However, this radiating surface picks up considerable energy radiated by the nearby earth, the amount depending on the altitude and orientation of the station, and on how much of the earth which can be seen from the station is lit by the sun. The actual temperature of this radiator will probably average about −30°F.

To keep the greenhouse temperature from rising above 90°F. it will have to be cooled by pipes filled with running water. To keep the temperature of the surface of these pipes no higher than 85°F. the heat from them will have to be removed by using this heat to boil ammonia. The ammonia vapor will be expanded through a turbine and condensed to a liquid again by contact with pipes in which liquid butane will flow; butane which has been cooled to −30°F. in the radiating surface. The turbine will drive an electric generator which will produce about 1,800 kilowatts. This is enough to supply light, air-conditioning, and communications and leave about 14 kilowatts per worker actually on the job for manufacturing.

Now for an estimate of how much all this will cost. We begin with the cost of the rocket fuel to lift the needed material into orbit.

The Apollo 15 rocket[3] weighed 6,407,758 pounds at lift-off. Of this, 309,330 pounds were put in earth orbit. Let us assume that a similar rocket is used in the shuttle service to bring up materials for building space stations. About 85 percent of the weight placed in orbit would be payload and 15 percent air frame and a minimal amount of fuel for retro-fire and approach to a conventional landing. This leaves 262,931 pounds of pay-load.

Now we need to know how much fuel was used to lift this payload into orbit and what this fuel would cost in large volume. At lift-off this rocket contained 437,650 gallons of liquid oxygen. At 9.5 pounds per gallon this amounted to 4,157,675 pounds of oxygen. It also contained 329,500 gallons of liquid hydrogen. At .58 pounds per gallon this amounted to 191,110 pounds. The third stage contained 20,150 gallons of liquid hydrogen (included in the above figure) of which only a fourth had burned when the rocket went into earth orbit (in this particular flight the rest was used to carry the rocket to the moon and then back to earth). Thus the packet in orbit must have contained 15,113 gallons of liquid hydrogen or 8,765 pounds. This is the stoichiometric equivalent of 69,563 pounds of oxygen which must also have been in this package. Subtracting these weights from the gross weight at lift-off one finds that 4,088,112 pounds of oxygen and 182,345 pounds of hydrogen were burned in lifting the payload into orbit. This hydrogen is the equivalent of 1,477,183 pounds of oxygen, leaving 2,610,929 pounds of oxygen still unaccounted for. This must have been burned with a hydrocarbon fuel. Assume that this fuel had the average composition $CH_{1.9}$, 2,610,929 pounds of oxygen would burn 769,837 pounds of the hydrocarbon completely to water vapor and carbon dioxide. There should, therefore, have been about this much on board at lift-off, and this was burned to get the payload into orbit.

Thus, to get 262,931 pounds of payload into orbit required 4,088,112 pounds of oxygen, 769,837 pounds of hydrocarbon, and 182,345 pounds of hydrogen. A practical shuttle rocket would require somewhat more. The booster and second stages of the Apollo 15 rocket fell into the sea and were lost. In the case of a practical shuttle rocket these would have to be equipped with wings and landing gear and land on a spaceport like conventional airplanes, ready to go again. This will require extra weight, and more fuel to lift this extra weight. In this estimate I make an allowance of 20 percent of the total fuel to provide this. Thus, in a practical shuttle one would need per pound of payload lifted into orbit, 18.7 pounds of oxygen, 3.51 pounds of hydrocarbon, and .83 pounds of hydrogen.

156

Now for estimates of the cost of these fuel items. I shall figure liquid oxygen at twice the cost, at 3¢ per kilowatt hour, of the theoretical minimum amount of power required to liquefy it from air, assuming heat interchange between the cold, unliquefied nitrogen and the incoming air. This works out to .97¢ per pound.

For the cost of hydrocarbon fuel in bulk, I use a figure three fourths of the present cost of fuel oil in my area, which is 40¢ a gallon, and thus 30¢ a gallon wholesale, or $12.60 a barrel. With 8 pounds of fuel oil per gallon this is 3.75¢ a pound.

Hydrogen will probably be made by the reaction of steam on red-hot coke, which requires little additional energy. Assuming that the coke used contains 80 percent carbon and costs $50 per ton, this contributes 19¢ to the cost per pound of hydrogen (the comparable cost of hydrogen from electrolysis of water at 3 volts per cell and 3¢ per kilowatt hour is $1.22 per pound). Double the materials cost of hydrogen gas from the watergas reaction, to take care of operating expenses, and you get 38¢ per pound. The minimum amount of energy required to liquefy one pound of hydrogen is 2.09 kilowatt hours, and at 3¢ per kilowatt hour this is 6.27¢ per pound. Double this to take care of possible inefficiency of the process and operating and administrative expenses and you get 13¢ per pound. This plus 38¢ per pound for hydrogen gas gives a total price of 51¢ per pound for liquid hydrogen.

Combining these figures one finds fuel costs, per pound of payload, of 18¢ for liquid oxygen, 14¢ for hydrocarbon fuel, and 42¢ for liquid hydrogen, for a total of 74¢ for the fuel to lift one pound of payload into orbit.

It is assumed here that the shuttle vehicles will be re-used, with preventive maintenance and replacement of worn parts, so that a vehicle is retired only when it becomes obsolete. By analogy with automobiles so treated I estimate that first cost, maintenance, and operating costs will amount to twice the fuel costs, giving a total cost of $2.22 per pound of payload put in orbit.[4]

We are now in position to reckon the cost of the station. The data are recapitulated in Table 1 below.

Table 1

Item	Weight in place pounds	Weight brought up, pounds	Cost, fob shuttle dollars per pound	Total Cost dollars
Granite	0	4,200,000	.01	42,000
Pressure hull	175,000	from granite		
Mirror	5,000	from granite		
Radiator	100,000	from granite		
Bulkheads	20,000	from granite		
Machinery	250,000	from granite		
Glass	80,000	from granite		
Oxygen	140,000	from granite		
Ammonia	10,000	10,000	.03	300
Sea water		200,000	.005	1,000
Salt	3,000	from sea water		
Sulfur	500	from sea water		
Water	180,000	from sea water		
Carbon	30,000	30,000	.03	900
Phosphorus	5,000	5,000	1.00	5,000
Copper	5,000	5,000	.60	3,000
Ti, Cr, Ni, Co, Mo, W As, Ga, Se,	2,000	2,000	5.00	10,000
Te, Rare earths	500	500	20.00	10,000
Au, Pt, Ag	100	100	1000.00	100,000
Totals	1,006,100	4,452,000		172,200

Here, as elsewhere in this book, dollars are of vintage 1974. It is assumed that as inflation or deflation occurs the values will go up or down together with relative values unchanged. Costs include only what we on earth will have to supply. It is assumed that we will run the shuttle service.

Multiplying the total amount which has to be lifted to establish this station by 2.22, the estimated cost per pound for lifting into orbit, one finds the costs of lifting to be $9,884,772. Adding the cost of materials to be supplied fob shuttle one gets $10,056,972 for the cost of the station. Dividing now by 250, we get $40,227 as the cost of the station per family. I realize, of course, that this figure is uncertain by at least a factor of two, up or down, and include figures beyond the first digit only so that the reader may easily check the details of the accounting.

This figure may now be compared with the cost of structures now required here in the United States to establish a new family of four. Table II, below, shows how I estimate these.

Table II

Estimated Cost of Structures Required to Establish a
New Family of Four in the United States

House	$20,000
Share of structures in which employed	20,000
Share of school buildings	1,000
Share of municipal facilities	500
Share of buildings and equipment providing utilities	2,000
Share of roads	1,000
Share of retailing facilities	1,000
Share of recreational facilities	500
Share of churches	200
Total	46,200

This figure is very uncertain, depending, as it does, on conventions of accounting. The most that can be had by comparing these estimates of the cost of establishing a new family in space or in the United States is that they are probably not very different.

In thinking about these figures it must be remembered that the cost estimated here for establishing a new family in space applies, not to the average cost, but to the cost of establishing one more family in space after tens of thousands of such families have already been established there, and the experience gained with such operations and the economies of mass buying, have been fully exploited. To establish the first family in space will cost at least ten times this, and the thousandth family perhaps twice this.

The social values accruing from the ability to put up colonies which will not interfere with other colonies may override the economic advantages in some cases. Fragments of ethnic groups so established will be able to preserve their own language and culture while able to communicate and trade with the rest of the world without encroaching on any other group's territory or livelihood.

Notes

1. These flywheels were mentioned in Chapter 3. What one has to do is to conserve angular momentum about two axes at right angles to each other. The motion has a degree of freedom, in that rotation about the polar axis does not affect the operation of the station until the speed of this rotation induces a perceptible centrifugal force. If the rate is one complete rotation per hour, the centrifugal force is only 10^{-5}g and would not be noticed.

The use of small flywheels within the hull, as contemplated here, may not be the most economical way of conserving angular momentum. I have not been able to work the problem through completely but it appears that if the wheels are too small and consequently have to rotate too fast, gyroscopic effects may complicate the use, and that above a certain critical angular velocity of the wheels control by this means alone may be impossible. The flywheel system may have to be supplemented, or even entirely superseded, by a set of moving weights just inside the pressure hull, and moving all around its periphery in two planes, or even by a set of weights on masts extending well outside the pressure hull. It appears that there are certain conditions between the moments of inertia of the station and the control means in order to have both conservation of angular momentum and efficient control of attitude, but I do not know what these conditions are.

Note that conservation of energy is not required, in that the system is constantly receiving energy from the sun and rejecting energy from the radiating surfaces. As a practical matter, however, there may be an upper limit to the amount of kinetic energy which can be stored in the flywheels if control of attitude by means of them is to be adequate.

2. It is interesting to compare the photosynthetic yields predicted here with those actually obtained in conventional agriculture. It is implied that the station produces about 400 watts of edible food energy per person; this amounts to .91 watts per square foot of projected area exposed to the sun. About the best that has ever been done terrestrially is 48,000 pounds per hectare of the new high-yield rice in regions where three rice crops are harvested per year. This amounts to .12 watts per square foot. The lower terrestrial yield is accounted for by (a) no production at night (b) production on cloudy days is only about a third of that on bright days (c) during the harvest period sunlight is wasted on bare ground (d) in the early growth period of rice much sunlight is lost by reflection from water (e) sunlight comes in obliquely much of the day. In our own Mid-west, raising corn and wheat, it is doubtful that we do even a fourth this well.

3. Letter to the author from Mr. William J. O'Donnell, NASA

4. It can be argued, on the ground that the analogy with automobile operations is defective, that the actual costs of getting a pound of material into orbit will be much higher than this even after tens of thousands of shuttle flights have been made and the experience gained in these used to perfect the vehicles and procedures. Although I have not heard or seen in print any estimate of what NASA expects the first shuttle

flights to cost, I gather that they are thinking in terms of more than a hundred dollars to put one pound of material in orbit. My explanation of this discrepancy is given in appendix 5. The argument against the validity of the analogy runs that our present rockets are made of materials more expensive than those which go into automobiles, that the cost of fabricating these materials is also larger, and that the electric equipment which goes into rockets is more sophisticated than that which goes into automobiles. Since we have as yet no operating experience with shuttle flights on which to draw, we cannot be sure which point of view is right but in rebuttal I offer the following: The shuttle rockets will be unmanned, with all guidance supplied first from the ground and then from the receiving station. The payloads, of the order of hundreds of tons, will be larger than those of present-day automotive vehicles, with corresponding economies. There will be no taxes or insurance. The above may contribute some plausibility, but the clinching argument is that for a wide variety of devices—guns, electric motors, airplanes, and the like the cost of the device is of the same order of magnitude as the cost of the energy it consumes during its useful life.

Appendix 2
Design of Interstellar Ships

The cost of interstellar ships depends almost entirely on the weight of material which arrives in the neighborhood of a distant star. This material was brought up to a tenth the speed of light at the beginning of the voyage and then slowed to nearly zero speed on arrival. We begin with design of the structures needed on actual arrival at a planet of the distant star.

In designing orbital dwellings we had to restrict ourselves practically to granite and sea water as raw materials because these are the only materials of which we have enough to support the scale of construction contemplated. This same limitation applies to the first stages of interstellar ships but not to the final stage. Here the total weight is small compared with the resources of the earth even if we launch ten thousand or more of them. At the end of the voyage it is weight and weight alone which is expensive; we can and should design in any material, no matter how expensive it is on earth, if by so doing we can make a significant reduction of weight on arrival. We must remember, too, that ship and rocket carcass will be salvaged three or four hundred years after the landing, at a time when the colony may be very short of rare metals.

We begin with the landing craft. We cannot know in advance what the conditions of the landing will be; perhaps on a planet with half again the gravity of earth — if it is much more than that the landing will fail because the colonists will never be able to get up and walk—or perhaps on a planet where gravity is only one tenth that of earth—if it is much less than that the landing will fail because the planet will not have been able to hold enough atmosphere. The density of the planet atmosphere may be three or four times that of earth—in which case design of the parachutes will be easy—or a tenth that of earth — if it is less than that, even it if is pure oxygen, it will not support human life. The most difficult problem which the landing craft can encounter and survive will be a high-gravity planet with a tenuous atmosphere. This combination will place the heaviest burden on the parachutes. These have to be designed and constructed after the colonists have been in orbit about the planet for some time and have sent a probe down to measure the planet's gravity and the pressure and composition of its atmosphere.

I do not see any way, either by changes in design or the use of more expensive materials, to improve significantly on the craft used to bring our lunar explorers back to earth. The capsule from the interstellar ship will come down on land, as the Apollo capsule is designed to do if necessary. Much of the electronic equipment which it contained will be unnecessary, and a smaller weight of batteries will suffice. The Apollo capsule carried only three people. This one will have to carry eight, but the crew of the Apollo capsule had to live in it for three days whereas interstellar colonists will be in it for at most two hours. More crowding can be tolerated and much of the life support equipment which was used in the Apollo missions can be left out. Against this, the interstellar colonists will have to carry camping equipment and supplies which the Apollo astronauts did not need.

The weight of the extra equipment which the interstellar landing craft will have to carry may be made up somewhat as follows:

People	8 @ 150 pounds[1]	1,200 pounds
Food	200 pounds each for 8 people	1,600
Water	to last until they can reach a stream	200
Scientific test kit for analysing soil, water, and plants		200
Medical kit		100

Tent and cooking utensils		300
Clothing	mostly boots; some winter gear	400
Microfilmed information and reading-equipment		20
Radio transceiver	for communicating with parent ship	20
Batteries for radio	rechargable	50
Hand-cranked generator for charging batteries		40
Plant seeds	many kinds	50
Agricultural tools		400
Total		4,580

The weight of the Apollo series command modules was about 11,700 pounds[2] and its diameter was 12 feet 10 inches. Stripped down it probably weighed about 10,900 pounds. We can cram the landing party and their supplies into it for the brief time they will be there. To do so makes the gross weight of the landing craft 15,480 pounds when it enters the planet's atmosphere, implying a loading per square foot 32 percent greater than in the Apollo landings. This may seem prohibitive until it is remembered that the Apollo craft plunged into the atmosphere at almost escape velocity, coming, as it did, directly from the moon, whereas the interstellar landing craft will come in at orbital velocity, which is 30 percent less.

Besides the main landing craft we shall need four smaller ones to be used after the main landing, to carry down more food, men, or animals, as the situation on the ground may require. These vehicles will no doubt be similar to those used in the Mercury series in the early days of space exploration. They will probably weigh about 2,500 pounds each, empty.

Also needed will be a one-man vehicle for outside work immediately around the orbiting ship. This vehicle will have its own propulsion system, view plates, and external claws which it will use in modifying the structure in orbit from its configuration during the deceleration phase to one very like the orbiting dwellings described in Chapter 3. This will weigh about 500 pounds. The probe which must be sent down before the landing will also weigh about 500 pounds.

Also needed will be a nuclear reactor to supply power to the station in orbit during the transition period (which may be as much as ten years) between the end of the deceleration phase, when the thermonuclear engine will be cut adrift and can no longer supply power, and the time when the orbiting station can get all the power it needs from the star about which the planet orbits. This reactor, developing a peak power of about 50 kilowatts, will be unshielded, and will have to be deployed at a distance of about five miles in order to keep the fast neutron flux in the station down to 20 per square centimeter per second.[3] The people inside can stand this without impairment of their health. Note that we do not try to keep the nuclear radiation flux in this vehicle as low as during the deceleration phase because we are no longer so concerned with genetic damage to the crew remaining in orbit. The genetic material banks in this ship, which will occupy little space, will have additional shielding. The biological effects of gamma radiation from this reactor under these conditions will be small compared with those from fast neutrons.

Very likely potassium will be used as the working fluid for converting the heat generated in the reactor into electrical energy. The heat rejected in the process will be radiated from black metallic vanes at about 500°F. Power will be generated in a very large but light electrostatic generator insulated by the vacuum of space, and transmitted to the orbiting station at high voltage on wires. These wires can be light and thin, for only tidal forces develop between the station and the reactor. The wires will also carry the signals which control the reactor, which is provided with means for ejecting spent fuel elements and replacing them from its supply of new ones. I estimate the total weight of this reactor plus connecting wires and control mechanisms at 2,000 pounds.

The weight of the ship used during the deceleration phase will be determined largely by the space which the crew will need for health and comfort. To review its mission: during the

166

deceleration phase, which will last about two years, about twenty-five people will live in it. These were born and grew up under conditions of weightlessness; during their life in this station they will be under a gravity of about 5 percent that of earth, and this will be uncomfortable and inconvenient for them, but it will help the landing party to prepare for life under the more severe gravity of the planet. At the end of the deceleration phase, after the landing party is away, the mission of the station will be to relay messages to and from earth, and to send down animals, food, or people, as needed on the ground, to the limit of its ability. An essential feature of this station is the garden, which operates under artificial light and supplies the food and oxygen required by the crew, all through the acceleration, coast, and deceleration phases, and after the landing party has gone.

During the coast phase, lasting perhaps several hundred years, the people lived under zero gravity in ample quarters; these can be large because they are jettisoned at the end of the coast phase and do not contribute to the final cost-determining weight. These ample quarters will contain neither greenhouse nor garden; they will depend for food and oxygen on the garden which will ultimately go into orbit about the distant planet. After the landing party has gone this station will gradually be rebuilt from within into a structure with an adequate greenhouse generating food, oxygen, and electric power from the light of the host star. In planning the vehicle to be used during the deceleration phase, provision must be made to ease this transition and bring it within the physical and mechanical means of the people remaining after the landing party has gone. For instance, panes of glass, which would be difficult to make on board, must be carried in sufficient number to equip the new greenhouse, with spares estimated to last a thousand years.

On the way out communication with earth will be by means of successive radar dishes carried at the end of the string of propulsion rockets and discarded, one after another, as each spent rocket is discarded. On completion of the deceleration phase, after the last rocket carcass has been jettisoned, the last radar dish remains rigidly attached to the orbiting

167

vehicle. One of the first things to be done after the landing party is away is to detach this antenna and place it in an orbit near the station, where it will be free to rotate and track the earth independently, leaving the station free to rotate so as always to have its greenhouse facing the star.

We would like to make the pressure hull of this ship spherical to minimize the weight per unit volume, but this is impractical. We need a bumper out in front to take the impact of the stray atoms and specks of rock in space and it turns out that this bumper must be thick and heavy. We have to keep the cross-section of the ship down to minimize the weight of this bumber, and also to minimize the chance of impact with meteors. In fact, we have to make the ship almost needle-shaped. We pay for the reduction of the weight of the bumper in terms of greater weight for the pressure hull. There is a limit to how needle-shaped we can make the ship; if it is too narrow we cannot use the space inside effectively as living and working quarters. To make the diameter the same as that of the landing craft, i.e., 12 feet 10 inches, seems a workable compromise. Now we have to see how long to make it, which is determined by how much space we need inside.

Before getting down to specifics there are some general features of the design which we have to settle. We know that shielding the twenty-five people on board from the nuclear radiation generated by impact with the stray atoms in space will be a serious problem, so we want to stow cargo, inside and outside the pressure hull, in the forward part, but we have to have access to every part of the inside surface of the pressure hull, without moving too much cargo, in order to fix any leaks which may develop. People will be continually moving around inside and must not get in each other's way, so we provide two ladder trunks, one for people going up and one for people going down, and also a trunk for ducts and wiring. We remember that during deceleration people will seem to weigh 5 percent as much as they do on earth, and cannot get around by swimming through the air, as they would in an orbiting station, so we provide flat surfaces for them to walk and sleep on. These same people will be living in this same hull under conditions of weightlessness for perhaps as much as two years

168

after the end of deceleration, so everything must be designed to be serviceable under conditions of weightlessness. Then, after the landing party has gone, the ship must be reconstructed from within to subsist on light from the host star. Finally we must remember that if the landing succeeds this ship will eventually be salvaged by a civilization very short of some metals. They will use these metals, so far as compatible with minimum weight, in the construction.

Now to the actual listing of the space required: We begin with individual staterooms, twenty-four of them. We can accommodate four per deck, and they need a ceiling height of 6½ feet. One entire deck will be for the Captain's cabin. We have to give up the idea of a wardroom and substitute three decks for general social use. The galley requires another deck, and we need two more decks for heads. During the deceleration phase a shower, a tub, and toilets will work; the shower can be re-rigged for use under conditions of weightlessness, but there will have to be special toilets for use after the deceleration phase. One deck needs to be set aside as an exercise room and another as a classroom. A command post will take up one deck, next to the Captain's cabin. The garden will require twelve decks. Five more decks will be needed for machinery spaces and cargo stowage. In addition, each stateroom and common room will have cubicles in which there will be either machinery or cargo. Thus we need a cylinder 208 feet long, with hemispherical domes which will add another 12 feet 10 inches, for a total length of 220 feet 10 inches.

The surface area of the pressure hull will be 8,903 square feet. It will be made of a stainless steel in which cobalt replaces part of the iron, which will be worked at a tensile stress of 50,000 pounds per square inch. With three pounds per square inch internal pressure the thickness of the pressure hull must be .00465 inches and its total volume 3.45 cubic feet. At a weight of 537 pounds per cubic foot this amounts to 1,852 pounds, to which must be added a like amount for bulkheads, for a total of 3,704 pounds for the pressure hull.

With the interior of the pressure hull at 85° F. and 70 percent relative humidity, the contained air will weigh 424 pounds.

The weight of the contents of the pressure hull is made up as follows:

Oxygen atmosphere		424 pounds
Liquid oxygen reserve	This is in liquid form, kept cold by a refrigerator whose weight is included in the estimate for machinery	4,000
Water	Includes the water contained in the plants and in food stores, but does not include the water content of the people	20,000
Food reserves	dry weight	6,000
Plants in the garden	dry weight	6,000
Food nutrient reserve	solids	500
Clothing and bedding	excludes those sent with landing party	1,000
Medical stores	drugs-only active ingredients, not conventional pills. Does not include medical stores for landing party	1,000
Sound and television	for entertainment; no books, all reading from TV screen	1,000
Library	on tapes, some video	2,000
Glass windows	for use in greenhouse at end of voyage	4,000
Other parts	for conversion at end of voyage	10,000
Jigs and tools for conversion		2,000
Total		57,924 pounds

The question how to protect the decelerated vehicle and the people in it from the effects of impact with stray atoms met in interstellar space has to be carefully thought out. There are three effects: a sort of sand-blasting in which part of the bumper is worn away; production of X-rays by the impact of the bumper with hydrogen atoms or molecules (practically it makes no difference whether the hydrogen atoms are joined or not); and production of gamma rays and particle showers by impact with stray heavy atoms such as carbon or oxygen. The impact of hydrogen atoms on the bumper at one tenth the speed of light is insufficient to split atoms, and so gives rise only to X-rays, but the impact of heavier atoms on the materials of the bumper can split the atoms of the bumper materials giving rise to a variety of penetrating radiations.

During the acceleration and coast phases the ample quarters used during the coast phase ride ahead of the smaller vehicle used during the deceleration phase and shield it and its crew from all these effects. The occupants of the ample quarters have to be protected from the radiation produced by impact by means of a heavy bumper, but since this is jettisoned at the end of the coast phase its weight does not enter the present calculation. At this point we are concerned with protecting the landing party, and the people who will later crew the orbiting station, from radiation during the two years of the deceleration phase, during which the speed of the vehicle drops from a tenth that of light to nearly zero.

According to Dr. Edward Purcell (loc. cit., page 8) we may expect to find one hydrogen atom per cubic centimeter in interstellar space. We know that there are other kinds of atoms out there too[4] — nitrogen, carbon, oxygen, and sulfur — in unknown concentrations, probably much less than one per cubic centimeter. I have assumed here that the effect of these can, for the purpose of this calculation, be lumped as one oxygen atom per 100 cubic centimeters.

In order to shield the sides of the ship from impinging atoms and meteorites during the slight yaws which occur as people walk around inside we make the bumper a little wider than the hull, say a radius of 6½ feet. The bumper is made of lithium-6-deuteride; lithium-6 rather than natural lithium because the absorption of neutrons in lithium-6 produces no penetrating radiations, whereas absorption of neutrons by natural lithium does, and deuterium rather than natural hydrogen for the same reason. To minimize production of X-rays by impinging electrons, and to get minimum radiation from atom-splitting by impinging heavy atoms we use, for the bumper, the elements of the lowest atomic number which combine to form a high-melting solid of good mechanical properties. These considerations lead inevitably to lithium-6-deuteride as the best material from which the bumper can be made.

The heat generated in this bumper by impinging hydrogen atoms is 274 watts. Impingement of heavy atoms adds 44 watts. The electrons associated with the hydrogen atoms are a current of 59 microamperes. For electrons, a speed one tenth that of light corresponds to a voltage of 2,400. Thus, the X-rays we can expect from hydrogen atoms striking the lithium-6-deuteride bumper are less than what we would get from a conventional X-ray tube operating at 59 microamperes and 2,400 volts. These are completely negligible physiologically. Any TV set generates more X-rays than this. The associated protons are not sufficiently energetic to go through the shield or to generate any penetrating radiation by impact with the lithium-6 or deuterium atoms. It remains to calculate the sand-blasting effect. In lithium-6-deuteride a proton coming in at a tenth the speed of light loses about 5×10^{-16} ergs per Angstrom unit. The heat of vaporization and dissociation of lithium deuteride is not less than 4×10^{-12} ergs per molecule. Thus it is difficult to see how the impact can, on the average, knock out even one molecule of lithium deuteride per impinging hydrogen atom. If one molecule of lithium deuteride were knocked out per impact during the two years of the deceleration period this would amount to about a millionth of an inch. The amount of sand-blasting by heavy atoms is difficult to calculate but it could hardly amount to more than 1,000,000 molecules of lithium deuteride per impact, or about .2 inches during the entire deceleration period.

Now to calculate the amount of exposure to nuclear radiation which we can permit in crew quarters during the deceleration phase. This will determine the weight of the bumper. According to Hanson Blatz[5] the radiation received from natural sources by people not exposed to X-rays or radium, before the atomic age, was about .00072 Roentgens per day. Of this about 20 percent was due to cosmic rays. Here on earth we are shielded from cosmic rays by the atmosphere, and in interstellar space they will amount to more, but not as much as ten times more, than we get here. To shield our colonists from cosmic rays would be prohibitively expensive; they will just have to take it, and their background exposure may be as high as .002 Roentgens per day. This may be compared with .042 Roentgens per day, which the Atomic Energy Commission considers safe. Since we are concerned here with the genetic material on which a whole new culture will be based, we must limit the amount of radiation which the crew receives from the impact of the bumper with stray atoms and from the operation of the nuclear engine during the deceleration phase, to no more than they get from natural sources and cosmic rays, say a total dose of 30 Roentgens up to the end of child-bearing. Since genetic damage is not repaired during an individual's life-time it is the total dose rather than the dose rate which matters.

The production of nuclear radiation by impact with stray atoms during the deceleration phase is proportional to the cube of the speed of the vehicle. It follows from this that the total dose[6] received by the crew during this period is one fourth the product of the initial dose rate in Roentgens per day by the duration of the deceleration phase in days. From this it follows that we can allow an initial dose rate as high as .0822 Roentgens per day from impingement with stray atoms.

In the absence of shielding, one finds the dose rate of gamma radiation by multiplying the production of gamma rays at a point source in watts by 19,500 and dividing by the square of the distance in feet from the source to the point where the radiation is absorbed. Let us assume that the nearest deck not used for cargo stowage is 100 feet from the center of the bumper, and assume further that of the 44 watts of impingement of heavy atoms one half goes into thermal effects, one fourth into gamma rays, and one fourth into fast neutrons. Then we have 11 watts of gamma rays at the beginning of the deceleration phase, and in the absence of shielding this implies a dose rate on this deck of 22 Roentgens per day. This must be reduced, by shielding, to .0822, which seems to imply a shielding factor of 261, but we must look a little closer. In shielding from gamma rays three processes are operative: photoelectric effect, pair production, and Compton scattering. Because in this case the target of the radiation subtends an angle of only 8° at the source, radiation scattered by photoelectric effect of pair production is likely to miss the target altogether, and because the shield is made of materials of such low atomic number, scattering of radiation by photoelectric effect or pair production is unlikely in any case. Thus we are concerned only with Compton scatter, but this process implies a build-up factor, since the incident photon is not destroyed, as implied by an exponential shielding law, but only deflected with loss of energy. In the case, usually considered, of an infinite plane wave of incoming radiation and a two-dimensionally infinite slab of shielding material, the build-up factor corresponding to a shielding factor of several hundred is of the order of magnitude of 10. Because the geometry of this ship-shield combination does not approximate that of an infinite source and an infinite shield, most of the scattered radiation which ordinarily contributes to the build-up factor will miss the target. Accordingly, I here assume a build-up factor of 2 and a shielding factor of 522.

We can get some shielding by interposing cargo, carried inside and outside the pressure hull, between the crew and the source of radiation. The total amount of this cargo, spread over the area of a deck, amounts to 420 pounds per square foot. Of this we can interpose only half between the crew and the bumper because we need the other half of the cargo in the stern to interpose between the crew and the radiation from the nuclear rocket engine. The shielding factor for gamma rays corresponding to a loading of 210 pounds per square foot is about 50. Now the shielding factor for a composite structure is the product of the shielding factors for the individual components of the shield, hence we need a shielding factor for gamma rays of 10.4 from the bumper.

Attenuation of gamma rays by the Compton process is proportional to the number of electrons per unit volume in the shielding material. For lithium-6-deuteride, with a density of .82 grams per cubic centimeter, this is .41 gram formula weights of electrons per cubic centimeter. For aluminum the corresponding figure is .13. It is known[7] that the e-folding distance for Compton scatter in aluminum is 25 centimeters at a gamma ray energy of 10 mev, and is insensitive to energy. Thus the e-folding distance in lithium-6-deuteride is 3.1 inches and the thickness of lithium-6-deuteride to get a shielding factor of 10.4 is 7.3 inches.

We must now find out whether a lithium-6-deuteride shield 7.3 inches thick is sufficient to protect the crew from the fast neutrons produced by the impact of stray heavy atoms on the shield. We have already found that this is about 11 watts. To find the Roentgens of fast neutron radiation per watt we divide 94,600 by the square of the distance in feet. Cranking in 11 watts this gives 104 Roentgens per day in the absence of shielding. Thus we need a shielding factor of 1,265.

With neutrons there is no build-up factor but they are known to be about 10 times as destructive to genetic tissue, erg-for-erg, as gamma rays. So, for practical purposes, we need a shielding factor of 12,650.

In the specific geometry of this ship a neutron, once deflected, has little chance of getting into a crew compartment. The resonance absorption of neutrons in the intermediate energy range by lithium-6 reduces the chance of this happening nearly to zero. We are concerned, therefore, only with neutrons which go completely through the shielding undeflected.

The cargo used in shielding is a composite of carbon compounds, glass, and metal. I calculate an e-folding length of 6 inches for this material, corresponding to a shielding factor of 7.4, leaving a factor of 1,710 which must be provided by the bumper. Taking the total microscopic cross-section of lithium-6 plus deuterium for scattering and absorption as 1.5

175

barns, the e-folding distance comes out 4.25 inches and the thickness required to give a shielding factor of 1,710 is 31.6 inches. Thus a bumper adequate to deal with gamma rays is inadequate for neutrons, and we have to use the thicker one. The volume of the bumper is thus 350 cubic feet. Lithium-6-deuteride weighs 52 pounds per cubic foot, for a total weight of 18,200 pounds. To this must be added about 5 percent for supporting structures, giving a total bumper weight of 19,110 pounds.

Insofar as nuclear radiation is concerned, the ship might survive a hit on the bumper by a meteorite as large as .14 inches in diameter (assuming it spherical) but would probably be destroyed by the mechanical effects of impact with a meteorite only .014 inches in diameter. The calculation follows: an instantaneous dose of 600 Roentgens would probably kill most of the crew, though all of them would live a few weeks after receiving it. 600 Roentgens corresponds numerically to a dose of .0822 Roentgens per day for 7,300 days or 6.307×10^8 seconds. As seen above, it takes 44 watts to produce .0822 Roentgens per day behind the shielding. Thus the total watt-seconds to produce a total dose, in crew quarters, of 600 Roentgens comes out to be 2.775×10^{10} joules or 2.775×10^{17} ergs. A particle weighing 61.6 milligrams at a speed of one tenth that of light has this much kinetic energy. If the particle is made of rock it has a volume of .0228 cubic centimeters and if it is spherical it has a diameter of .14 inch. This is the largest meteorite which the crew could survive if the ship hit it just as the deceleration began. A year later it could survive impact with a meteorite of twice that diameter.

The kinetic energy associated with a meteorite .14 inches in diameter at a speed a tenth that of light is the same as the total energy released in the explosion, in air, of 9,000 pounds of nitroglycerine. By terrestrial standards our space ship is a fragile structure. At the distance the hull is from the bumper, it is doubtful that it could survive the explosion of one pound of nitroglycerine in air. However, in the vacuum of space, and at the very high energy density incident to impact at a tenth speed of light, the explosive would, pound for pound, be less destructive. There would be less to transmit the shock, and a greater fraction of the energy would be dissipated as light and heat. Still, even in a vacuum and at high energy density, it seems doubtful that the ship could survive an explosion releasing the energy of ten pounds of nitroglycerine. This is why I think that a diameter of .014 inches describes the largest meteorite which the ship could hit and survive at the beginning of a deceleration period.

We are now in position to calculate the total weight of the ship just before the landing. It is made up as follows (to nearest fifty pounds).

25 people @ 130[8] pounds	3,250 pounds
Hull and fittings	40,000
Stores	47,500
Landing craft and probe	21,400
Equipment for use after landing	4,600
Repair vehicle	550
Nuclear reactor	2,000
Antenna	2,000
Bumper	19,100

Thus the total weight of ship, people, and landing craft and equipment is 140,400 pounds and the thrust required to get it down to zero speed in two years is 6,810 pounds. It now remains to design a nuclear rocket which will furnish this thrust for two years.

First, choice of fuel: we can make that by a process of elimination. We cannot use nuclear fission because the rocket would be too heavy. We cannot use either fusion of hydrogen to form helium, or the mutual annihilation of positive and negative matter because the necessary technology is too far in the future. We cannot use the reaction of deuterium on deuterium because that generates so many fast neutrons that we could not shield the crew against them. We cannot use the reaction of deuterium on tritium because tritium is radioactive with a half life of about ten years and would decay during the hundreds of years our fuel may have to be stored. This leaves the reaction of deuterium with helium-3. So far as is

177

known, this reaction produces no penetrating radiation. Deuterium and helium-3 are stable substances which can be stored indefinitely, though this will require some refrigeration, since the boiling point of helium-3 is below the temperatures which can be reached by radiation into space. Helium-3 can be made from deuterium, and there is enough deuterium in the sea to fuel more than a million interstellar ships.

We have now to compute the maximum rocket nozzle velocity we can get from this fuel. The overall reaction is deuterium plus helium-3 to give helium-4 plus hydrogen. According to Lapp and Andrews (loc. cit. Table 3-2, page 56) the sum, in units of atomic weight, of the helium-4 plus hydrogen isotopes is 5.012018 and the sum of the weights of the helium-3 plus deuterium isotopes is 5.031738. Thus, in the reaction, .019720 units of mass are lost per 5.012018 units out of the nozzle. Invoking the equivalence of mass and energy we find that the theoretical nozzle velocity is .088708 that of light. To get the effective ejection velocity from this we have to make two corrections. Our engine cannot be perfectly efficient. So far as thermodynamic limitations go it could be more than 99.9 percent efficient but we have to give our designer some scope, and so assume an efficiency of 99 percent. The 1 percent loss of power decreases the nozzle velocity to .088263. We cannot shoot directly aft because our jets would strike the fuel tanks and burn them up, so we angle the jets 2° outward. Thus we must multiply our corrected nozzle velocity by the cosine of 2° and find an effective jet velocity of .088209 that of light.

Now we have to compute the weights of the fuel, the rocket carcass, the fuel tanks, the connecting lines, and the radiation shield.

Although to get the space ship and associated apparatus down to near zero speed in two years requires a pull of only 6,810 pounds, the total thrust generated by the nuclear engine has to be about 53,900 pounds at the beginning of the deceleration period in order to slow the fuel and the rocket structure down too. At startup the power is about six billion kilowatts — about ten times as much as was generated by the

178

Apollo rockets at lift-off. It is assumed here that about 99 percent of that power is used in generating thrust, leaving about 60 million kilowatts to be radiated as heat. This will be done from the forward part of the surface of the rocket, which is made of tungsten sheet, and works at a temperature of 2,000° K., radiating 840 kilowatts per square foot. We do not use a higher working temperature because this would lead to excessive evaporation of tungsten into the vacuum of space. We make the diameter of the rocket carcass 12.8 feet, conforming to the size chosen for the space ship, and find that the high temperature secion must be 180 feet long. Behind this is another radiating section 100 feet long operating at 400° K. We need to radiate at this temperature to keep parts of the rocket in which electric insulation is required, cool enough not to burn this insulation, and to radiate heat rejected by the refrigerating devices which keep the rocket magnets cool.

There are three of these magnets, each in the shape of half a toroid. The windings are refrigerated to the point where the metal becomes superconducting. Each end of each half-toroid terminates in a jet — six of them spaced 60° apart. The total mass flow through these jets, at the beginning of deceleration is .02 pounds per second. Part of the energy of the jets is recovered by a magneto-hydrodynamic process and used to accelerate deuterium atoms which are injected into the center of each half-toroid (note the similarity to an ordinary jet engine, in which part of the energy is extracted from the jet and used to run the compressor) where they strike helium-3 atoms and react, forming the jets.

The magnets are estimated to weigh 20,000 pounds each. The electrical machinery to drive the magnets, refrigerators, ion sources, and accelerators, is estimated to weigh 60,000 pounds. The pressure hull and internal fittings make up 40,000 pounds more for a total weight of rocket engine plus carcass of 160,000 pounds.

The reactions postulated produce no penetrating radiation but it would be naive to assume that nuclear reactions on this scale could run clean. Some neutrons and gamma rays will inevitably be produced and will radiate in all directions. The fuel interposed between the engine and the crew will shield the crew adequately from these, so long as as much as one tenth of the original fuel remains, but after that the crew will need protection. Besides that, the heat radiated as neutrons may be considerable, and the fuel tanks, which can stand only a few watts of heat, must be protected. For this reason a nuclear radiation shield is inserted between the engine and the fuel tanks. It is made of lithium-6-deuteride and weighs about 42,600 pounds.

The tank for holding the deuterium component of the fuel is 12.8 feet in diameter and 500 feet long. It weighs about 4,000 pounds. It contains an electric heater for vaporizing the fuel as needed by the engine. It runs at the equilibrium temperature of starlight, or about 10° K.

The tank holding the helium-3 component of the fuel is likewise 12.8 feet in diameter, 750 feet long, and weighs 7,000 pounds. The hydrostatic pressure at the bottom of this tank, due to the acceleration, is 1.2 pounds per square inch. To keep the vapor pressure below one pound per square inch this tank has to be run at about 1° K. It picks up 20 milliwatts of heat from starlight and 30 milliwatts by thermal conduction along the connecting cables. This heat has to be removed by a refrigerator which discharges heat at 20° K into a radiation surface 100 feet long below the helium-3 tank. This tank also contains an electric heater for vaporizing the fuel to the engine.

Connecting the engine, the nuclear radiation shield, the deuterium tank, the helium tank, and the space ship, there is a triad of cables in the form of hollow tubes. Besides transmitting the force which tows the tanks and space ship, these cables carry electric power from the engine to the other components of the string, and, serving as wave-guides, carry instructions from the computer in the space ship to the other components. The total weight of these cables is about 1,000 pounds.

We are now in position to compute how many kilowatts of neutrons the nuclear engine can generate without endangering the crew or their offspring—in other words, how clean this engine has to run. We have already made an allowance of .5

Roentgens total dose to the crew from this source. Cranking in the shielding factor from cargo in the after portion of the ship we find that we can stand a total dose of 3.7 Roentgens ahead of this shield.

The shield, made of lithium-6-deuteride between the engine and the deuterium storage tank is 72.5 inches thick and its shielding factor is 25 million. The dose we can stand forward of that shield is consequently 92.5 million.

Now to work out the shielding effect of the fuel. Assuming .5 barn cross-section for both deuterium and helium-3 the e-folding distance comes out .409 feet. The initial length of the column of fuel is 1,250 feet and it is all used in 730 days, so the length of the column decreases 1.712 feet per day, or 4.187 e-folding distances. Thus

$$D = \int_0^{730} R_o e^{-4.187 \, t} \, dt$$

where R_o is the rate, in Roentgens per day, of radiation which can be tolerated ahead of the previous shield, and the time, t, is in days. This gives R_o 4.187 D or R_o = 387 million Roentgens per day.

We have seen above that the Roentgens per day in the crew compartments, in the absence of shielding, is found by multiplying the watts of neutrons at the source by 96,400 and dividing by the square of the distance, in feet, from the source to the compartment. Putting this together with a rate of 387 Roentgens per day we find that a neutron output of 2.9×10^9 kilowatts at a distance of 8,500 feet, passing through all the shielding assumed here, would give the crew, during the whole of the deceleration period, a total dose of .5 Roentgens of fast neutrons. The neutron output of the engine cannot possibly be that high — 2.9×10^9 kilowatts is about half the peak output of the engine — so the crew is in no danger from this source. To find out how dirty the engine can run we have to use another criterion.

We note that unless the engine is running there are no neutrons, and if the engine is running we are evaporating deuterium from this tank at the rate of .008 pounds per second. This evaporation cools tank and fuel at the rate of 1.5

kilowatts. The tank is 6,000 feet from the engine, and we calculate that if, as computed above, the shielding factor of the lithium-6-deuteride disc is 25 million, the neutron flux from the engine can be as high as 10^{14} kilowatts. This is thousands of times the maximum output of the engine and we still do not have a criterion.

Next we look at the heat absorbed by the shield and require that its temperature stay below 100° C. At this temperature the shield can radiate 100 watts per square foot (remembering that only one side of it can radiate; we cannot allow radiation toward the tank). With this shield 4,000 feet from the engine, we can allow a total neutron production of 20 million kilowatts.

If the designers of the future can build the engine to run cleaner than that they can reduce the weight of this shield, but not very much, since the weight is proportional to the logarithm of the energy the shield has to absorb. If the actual engine cannot run that clean the shield will have to be modified, putting a graphite disc about a foot thick out in front, so that it can be allowed to radiate at any temperature up to 3,000° K, and radiate several thousand kilowatts per square foot, placing thin tungsten radiation shields between the graphite and the lithium-6-deuteride, and providing a skirt perhaps a hundred feet long to radiate heat absorbed in the lithium-6-deuteride.

We are now in position to calculate the total weight, rocket plus fuel and tanks, and space ship plus accessories, at the beginning of the deceleration phase. This weight is made up as follows:

Space ship plus accessories	140,400 pounds
Rocket carcass plus engine	160,000
Fuel	755,000
Cables	1,000
Deuterium tank	4,000
Helium-3 tank	7,000
Radiation shield	42,600
Total weight	1,110,000 pounds

To get the weight at the end of the acceleration phase we have to add to the above 200,000 pounds for ample living quarters, to be used, in addition to those already specified, during the coast phase, and 100,000 pounds for a lithium-6-deuteride bumper to ride ahead of these quarters. This may seem a skimpy bumper, but it must be remembered that the carcass and tanks of the last rocket from the acceleration phase ride ahead of this and perform most of the bumping function. These, with the ample quarters and associated bumper, are jettisoned at the beginning of the coast deceleration phase. The total weight at the end of the acceleration phase is thus 1,410,000 pounds.

We have now to calculate the fuel required to bring this mass up to a tenth the speed of light. Since in this operation energy, rather than weight, makes up most of the cost, we re-design the rockets to get the most out of the fuel rather than the most out of the weight.

A rocket operates most efficiently when the velocity of ejection is always kept equal to the forward velocity. Operating in this way the material ejected is left at rest and has associated with it neither kinetic energy nor momentum. We cannot realize this ideal exactly because (a) the highest exit velocity we can get from the deuterium-helium-3 fuel is only 88 percent of the final velocity, and (b) some parts of the rocket structure and all of the spent tanks must be jettisoned as we go along, at zero velocity relative to the vehicle, and thus these parts carry kinetic energy and momentum which must be accounted for in the calculation. For the purpose of the calculation I assume that the velocity of ejection is always equal to the forward velocity, and that for every pound of material making up the jet .15 pounds of structure is left behind. [9]

Invoking conservation of momentum:

$$mv + \sigma \int v \, dm = \text{a constant}$$

where m is the remaining mass of the rocket and v its speed, both functions of the time.

We sense logarithmic terms coming up and want our variables dimensionless, so we make the following transformations:

$$x = \frac{m}{M} \qquad\qquad y = \frac{v}{V}$$

where M is the final mass of the rocket and V its final velocity. Note that as we move toward the end of the flight, y increases from zero to one, while x starts out very large and decreases to one.

Conservation of momentum is now expressed by

$$xy + \sigma \int y \, dx = \text{a constant}$$

We want to get rid of the integral sign and so differentiate with respect to x, and have

$$x\frac{dy}{dx} + \sigma y = 0 \text{ or } \frac{dy}{y} = \frac{-(1 + \sigma)}{}\frac{dx}{x} \qquad .$$

This integrates to

$$\log y = -(1 + \sigma) \log x + \text{a constant}$$

Since $y = 1$ when $x = 1$ the constant is zero, and

$$y = x^{-(1 + \sigma)}$$

We now invoke conservation of energy and find

$$\tfrac{1}{2}MV^2 + \tfrac{1}{2}MV^2\sigma \int_{\infty}^{1} y^2 dx = \text{the total energy we must get}$$

from the fuel.

Substituting the above value of y in terms of x and noting that dx is negative we evaluate the integral and get energy from fuel needed for the acceleration phase

$$= \tfrac{1}{2}MV^2\left(\frac{1 + 3\sigma}{1 + 2\sigma}\right)$$

184

Taking σ as .15, which is representative of present experience with rockets, the factor is 1.115. We have found that M = 1,410,000 pounds and V is one tenth the velocity of light, and so are in position to calculate that the acceleration phase will require 2,000,000 pounds of fuel.

We are finally in position to reckon the cost of launching an interstellar ship. Heavy water now costs $55[10] a pound, corresponding to a price of $275 for deuterium. To this we add $1 for extraction from the heavy water and $10 for delivery in a remote orbit about the sun, for a total price of $286 a pound. We need 800,000 pounds of deuterium for acceleration, 302,000 for deceleration, and 15,000 in radiation shields, for a total of 1,117,000 pounds. This costs $319,500,000.

We need 1,200,000 pounds of helium-3 for acceleration and 453,000 for deceleration, for a total of 1,653,000 pounds. Helium-3 is made from deuterium; 4 pounds of deuterium give 3 pounds of helium-3. Thus the base cost of helium-3 is $368 a pound. To this we add $100 per pound for processing and $10 a pound for delivery, and get a total price of $478 per pound. Thus helium-3 will cost us $790,100,000.

We also need 45,000 pounds of lithium-6 at about $1,000 a pound — total cost $45,000,000.

We will need about 800,000 pounds of structures and machinery, quarters, rocket carcasses, and tanks. Some of these are made of expensive metals, and all of them have to be delivered to a remote solar orbit. The estimated cost is $40 a pound at a total cost of $32,000,000.

We shall also need about 50,000,000 pounds of hydrogen to slow down the rocket exhaust while the caravan is still at low speed. This is estimated to cost $15 per pound delivered at the remote solar orbit, for a total cost of $750,000,000. Thus the total cost of launching an interstellar ship comes out $1,936,000,000.

Many assumptions enter this figure which is therefore quite uncertain, but I think that any plausible calculation of this cost, based on the information now available, will be between one billion and five billion dollars per launch.

Notes

1. People in weightless space for generations will be lighter and smaller than the average person today.

2. Letter to author from Mr. William J. O'Donnell, NASA.

3. Hanson Blatz, *Introduction to Radiological Health*. New York: McGraw Hill, 1964, p. 114.

4. Dr. Ralph Becker, *Chemical and Engineering News*. June 3, 1974, p. 17.

5. *Introduction to Radiological Health*. New York: McGraw-Hill, 1964, p. 84, Table 1-12.

6. Let D be the total dose and R_o the initial dose rate. The instantaneous dose rate is $R = R_o(1 - \frac{t}{t_o})^3$ and $D = \int R dt = R_o \int_o^{t_o} (1 - \frac{t}{t_o}) dt$. Now set $u = \frac{t}{t_o}$ where t_o is the duration of the deceleration phase; then $dt = t_o du$ giving $D = R_o t_o \int_o^1 (1 - u)^3 du = 1/4 R_o t_o$.

7. Lapp and Andrews, *Nuclear Radiation Physics* New York: Prentice-Hall, 1959, Figure 5-13, p. 115.

8. We are averaging the weight of the ship occupants, including children.

9. The reader will have noted that to keep the jet speed at the designed value we have to mix inert materials, in this case hydrogen, into the jet.

The reader may also have noted, and been bothered by the fact that this calculation requires infinite mass at take-off. The limiting case is easily handled and does not complicate the mathematics. In the practical case we will build the caravan in orbit about the sun at a place where the orbital motion is toward the target star. Thus the initial velocity will not be zero and the initial mass not infinite.

10. *Federal Register*, April 18, 1974.

Appendix 3

Energy for the Space Program

If the space program continues somewhat along the lines suggested in Chapter 8 our total requirements in the United States may be about as follows:

Table I

Forecasted Annual Rate of Energy Consumption

	Space Program			Other		
Year	Fuel	Electricity	Total	Fuel	Electricity	Total
1976	.000 005	.000 003	.000 008	1,800	160	1,960
1978	.000 016	.000 009	.000 025	1,731	209	1,940
1980	.000 049	.000 028	.000 077	1,699	254	1,953
1985	.008 730	.004 890	.013 620	1,760	358	2,014
1990	.155	.070	.225	1,998	486	2,840
2000	49	28	77	2,994	695	3,640

In Table I the unit is one million kilowatts. In the case of electricity this is the actual forecasted rate. In the case of fuel, this is the amount of heat which can be had from the fuel used. The fraction of the heat produced by burning of fuel which can be converted into electricity varies from 15 percent to 60 percent according to the type of engine used.

Natural gas, petroleum, coal, and oil shale are substantially inter-convertible resources. Modern refining technology is such that a given weight of natural gas or petroleum, which are made up of carbon, hydrogen, plus a little sulfur, can be apportioned at will between gas for heating purposes, fuel oil, aircraft and rocket fuel, and gasoline. Coal and oil shale contain, in addition to carbon, hydrogen, and sulfur, nitrogen (useful as an essential constituent of fertilizers), oxygen, and metals, mostly calcium, magnesium, aluminum, iron, silicon, and rarer metals used in the electronics industry. From the standpoint of fuels, one pound of coal is the equivalent of about .6 pounds of petroleum, and one pound of oil shale to about .05 pounds of petroleum[1] (these factors take into account the fraction of the coal or oil shale products which are burned for heat to run the refining processes).

The percentages of gas, gasoline, jet, rocket, and diesel fuel, and heating oil produced from petroleum or natural gas during refining, are determined by the choice of temperatures, pressures, times, and catalysts used. Some lubricating oil may also be produced.

Liquid and gas fuel was first obtained from coal, more than a hundred years ago, by the process of coking, in which coal is heated, out of contact with air. The hard residue, coke, consisting mainly of carbon and ash, was used in blast furnaces for making pig iron. The gas from the coal was used for street lighting, cooking, and to a limited extent in internal combustion engines for making electricity. The distillate from the heated coal contained some ammonia, which was used in fertilizers, some hydrocarbons, which were used as fuel and for making chemicals, and some solid hydrocarbons which were used for making dyes and drugs.

About 1910, when Germany was faced with the prospect of a war during which petroleum supplies might be cut off, the coking process was taken a step farther. Hot coke was made to react with steam, which produced hydrogen and carbon monoxide; these two gases were then made to react, at a lower temperature and in the presence of catalysts, to produce a liquid quite like petroleum, and which is, for practical purposes, interchangeable with petroleum. This is the Fischer-Tropfsch process. It works very well. Between 1918 and 1940 Germany, Great Britain, the United States and Japan all produced petroleum substitutes from coal in pilot

188

plants, against the eventuality of a second world war. Germany captured the Russian oil fields, and Japan the Indonesian oil fields before their petroleum reserves ran out, and the United States was able to keep Britain supplied throughout World War II so it was never necessary to turn to coal to keep the military machines going. Meanwhile the United States Bureau of Mines, looking far into the future, ran a pilot plant for producing petroleum substitutes from oil shale.

Two points must be made completely clear: one, that coal and oil shale which we have in very large quantities in the United States are completely interchangeable with petroleum as fuel for rockets, vehicles and power plants; and two, that the technology for doing this is completely understood and demonstrated. The talk we hear about the need for research in these areas is just a smoke screen to delay matters until the financial problems associated with the switch from petroleum to coal and oil shale can be worked out. It is true that more laboratory work might improve existing processes by a few percent, but what we really need now is operating experience with large scale plants.

Here in the United States we have known for a long time that our petroleum reserves would soon run out. We have done nothing about it because cheap oil was available in the Middle East. So long as that was true it was uneconomical to turn to our own resources. Until there are some basic changes in our social outlook and tax programs we cannot do so even now, because if anyone did invest in a large scale program of production of liquid and gas fuels from coal and oil shale the Middle East nations would drop the prices of petroleum and bankrupt him. What we have to do is to rearrange our tax structure so that anyone who invests in plants to convert coal and oil shale into liquid and gas fuels is sure to make money no matter what the Middle East nations do.

This will be difficult for us because of the inevitable consequence that the rich will get richer while the poor get only a little richer. I am incensed at the thought that deserving people like you and me will be asked to tighten our belts and pay more taxes while undeserving people, whose only claim to merit is that they have money and are willing to risk it with the hope of having more money still, will actually become very rich indeed. Though I know better, I cannot refrain from writing my congressmen, inveighing against the inequity of malefactors of great wealth, but in my rational moments I realize that this is nonsense. The fact is that we are in a jam, and will have to buy our way out, and the only question is who can get us out at the least cost. There is no doubt at all that private industry can do the job at a cost lower than that of any other agency. We will just have to swallow our moral indignation and encourage them to do it.

It is often thought that we are a great nation because we are smarter and more industrious than anyone else. The fact is that we are not either as industrious or as good organizers as the Chinese, not as imaginative as the Arabs, not as smart as the British Indians, not as perceptive as the Blacks, and not as inventive as the Latins. What we do have is the ability to forego immediate advantage to accomplish long term goals, and the ability to compromise and work out a sound public policy by consensus. These attributes have given us long periods of stable government, during which great plans could be brought to fruition. We can rely on our government; we know that there will be no abrupt changes in policy, no unnecessary meddling in private business, and no capital levy. We know that our government will keep its promises even though these later become unpopular socially. This is the advantage which sets us above and apart from all others, and which will get us safely through the energy crisis, and make us a petroleum exporting nation again.

Airplanes and rockets will require petroleum-type liquid fuels for a long time to come, but for heating and roadbound vehicles we have an option. We can supplant our present heating and vehicle systems by electric types. Vehicles first: it is presently supposed that battery powered automobiles are impractical. Certainly they are not now available, though they outnumbered internal combustion powered cars about 1912. Most personal car driving is either commuting to work, by one way trips of twenty-five miles or less, or shopping trips, twenty-five miles or less there and back. If you take an ordinary one ton pickup truck, load in a thousand pounds of ordinary car batteries, take out the gasoline engine, and substitute a twenty horsepower electric motor for it, you have a serviceable commuting or shopping vehicle. If you commute, you recharge the batteries while you are at work; if shopping you recharge them when you get home. If they were assured of a market for them, automobile manufacturers could supply a better, longer range, and less expensive battery powered vehicle than the converted pickup truck visualized here. If we put a tax of a dollar a gallon on gasoline, exempted battery powered vehicles from all taxes, and reserved the main commuter routes for them exclusively during commuting hours, there would be a market for battery powered vehicles, and our need for petroleum could be cut nearly in half.[2]

As to household heating, the first cost in a new house, for thermal insulation, wiring and heaters, would be no more than that for a furnace and duct work. In older houses, or even in new ones in chilly areas, heat pumps could be used. An air conditioning unit run backward can generate up to five times as much heat as you can get by burning the same amount of power in an electric heater. Such units are commercially available. They are much used in Hong Kong.

Replacing internal combustion engines by battery driven motors, or oil-fired house heating plants by electric heaters, makes no sense, of course, if we are to continue, as we do now, to generate much of our electricity in oil fired plants; but if, as I think we will, we turn to nuclear power for almost all our needs for electricity, it will make a great deal of sense.

Before we commit ourselves to a nuclear power, fuel from coal and oil shale program, we should look around to see what other sources of energy may be available. These may be divided into two groups — those which, like nuclear power and power from coal and oil-fired plants, can operate continuously, twenty-four hours per day, and others, including power from solar heat, which are available only intermittently. Power sources in the second group are at a serious financial disadvantage, as we shall see.

Except for nuclear power and power from fossil fuels, the

191

first group, of sources which can operate continuously, contains only three: geothermal power, power from the mixing of salt water with fresh, and power from thermal gradients in the sea.

Power is available from heat deep in the ground. Virtually all of this comes from the decay of radioactive materials in the rocks. Near the surface of the ground the temperature increases about 2°F. per hundred feet as you go down. Geothermal energy has been used for many years in Italy and studies of its possible use have been made for California and Japan. Apparently the best way to use more of it would be to find a fissure in the rocks deep under ground — one several hundred miles long — and pump cold water into it at one end and get hot water under pressure out at the other — you do not let the water boil because this would require larger pipes to handle it — but rather use the heat from it to boil other water at the engine.

There are some open questions about further exploitation of geothermal power. If it is really so hot down there, why isn't ground water hotter than it is? And why is not the Humboldt River, which flows under the State of Nevada, hot? Apparently for efficient utilization of additional geothermal sources one must go very deep, and the rate of flow must be kept down, to give the water time to heat up. Perhaps we should have a small program to drill holes hunting for long natural fissures very deep down; but it would be unrealistic to count on geothermal power aside from what we now know about.

Power can be generated continuously from thermal gradients in the sea. As you descend into the water of the tropics there is little change of temperature until you get down to about 800 feet, at which depth the temperature of the water changes abruptly (within a range of twenty feet) from warm to cold. There is a change of density too; submarines can float in the thermocline. To utilize the difference in temperature to produce power one needs to run a pipe down to about a hundred feet below the thermocline and pump cold water up through it to the surface. Relatively little pumping power is required because the weight of the water in the pipe is almost balanced by the weight of the water outside. The pipe has to be well insulated thermally, otherwise the cold water coming up would be warmed by the water outside the pipe. At the power plant on shore the heat from warm surface water is used to boil a liquid (probably freon would be used); the vapor drives a turbine attached to an electric generator, and the vapor is condensed by heat exchange with the cold water from the pipe. From surface water at 75°F. and cold water at 34°F., a flow of one cubic foot per second in the pipe will get about 10 kilowatts per cubic feet of water per second pumped. Out of this about .4 kilowatts would be needed to pump the water. Such a plant, producing about 30 kilowatts, operated for several years near Matanzas, Cuba, during the 1920s.

A difficulty with this process is that nowhere that I know of, off the continental United States, is there a place from which the thermocline is accessible from the shore. It may be economical to build such plants at the edges of the reefs south of the Florida Keys. There are undoubtedly suitable places in the Bahamas and the Caribbean, and perhaps in the Hawaiian Islands.[3]

In theory, at least, power is available continuously from the mixing of fresh with salt water at the mouths of rivers, as Maria Telkes pointed out many years ago. Pairs of electrodes, one in salt and one in brackish water, develop an electrical potential difference of about .1 volt. There does not appear to be any useful way in which such electrochemical cells can be connected in series to give a higher voltage, but it is possible that by means of mammoth tunnel diodes this minute direct-current voltage could be converted into high-voltage 60 cycle current. The difficulty is that the electrodes would apparently have to be about a mile apart, and the currents involved would be hundreds of thousands of amperes — the wire bringing this current to the tunnel diodes would have to be very heavy. I do not think that power from the mixture of fresh with salt water is practical.

It might seem that water power from rivers should be included among those which operate continuously. Hydroelectric installations at Niagara and on the Colorado and

Columbia rivers do operate continuously but it is unlikely that any other power sources in this category can be developed in the United States. The new ones will all operate seasonally, producing power for weeks at a time and remaining idle at other times. I worked for some months near Hiwassee Dam, part of the Tennessee Valley Authority. The water wheel there spun idly for months at a time, running "condensing", as it was said, contributing, by means of its inertia, to the stability of the electric system to which it was connected, but consuming, not producing power, with no water running through it.

Water power from rivers was immensely important to us a hundred years ago but it is a relatively minor resource now — there is not much more in sight. Individual householders near a foaming brook may wish to make small installations to reduce their power bills. These can be completely automatic and unattended, but power companies don't like them because they reduce the stability of the power system as a whole.

One has only to visit Lowell, Massachusetts, to see how important river power once was; the Merrimack is lined with factory buildings which were the largest in the world when they were built. Behind these buildings there are canals which bring water from above the rapids to discharge through water wheels under the buildings. From such installations we have a technical legacy — 550 volt motors which are used in the textile industry and nowhere else, and 40 cycle electric power which was used there and nowhere else. The Swiss railways once ran on water power generated at 16 ⅔ cycles; Niagara generated power at 25 cycles, which made the lights flicker, and Michigan had 30 cycles, which was almost as bad. These are gone now, unregretted, but they are an important part of our technical history.

Power from the tides is potentially more important. The French have such an installation at Quiberon Bay which is doing well. Our first one may be in Passamaquoddy Bay, for which the studies, surveys, and engineering have all been done — all we are waiting for is the money. Much more power could be had from the tidal flow in and out of Minas Basin in Nova Scotia where the range between high and low tides — 50 feet — is the highest in the world (the power to be had from a given bay is roughly proportional to the square of the tidal range). The flow may be either into Chignecto Bay — dammed at its mouth and kept perpetually at low water — or into the Gulf of Saint Lawrence. Conservationists should visit this area — anything we do there will be an improvement.

Tidal heights are predictable, though they do depend somewhat on weather. It might seem at first that all one had to do would be to have two reservoirs and by means of gates keep one of them always at low tide level and the other at high tide level, and run a waterwheel 24 hours a day on the difference in head between them. This is true in the limiting case of very large reservoirs, but to do this between reservoirs of finite size one would have to restrict the flow through the waterwheel, and so the amount of power generated, to so small a value that it would not affect the height of the water in the reservoirs appreciably during a twelve hour period. From any actual system one can get more power by intermittent operation, using enough water to change the level in the reservoirs appreciably during a twelve hour period. The schedule for most economical operation depends on the slope of the sides of the reservoirs and the width and depth of the channel (the momentum as well as the height of the water is important) and it is difficult to come up with a useful generalization, but it appears that in many cases power should be generated for about two hours to either side of high tide and low tide but at a varying rate, from a peak at high and at low tide and dropping off to half just before the shut downs in the mid-tide region.

In some localities it may be practical to generate power from wave action, but so far as I know this has never been demonstrated on a kilowatt or larger scale.

Wind power has been used for a long time, first to drive ships, then to push pack animals along (sails are still set to short masts on the backs of llamas in the passes of the Andes) and finally in windmills. Wind power may have been used to mill grain even before water power. Conservationists will probably not object to windmills, if they are built on the Dutch model, but they should be warned that extensive use of windmills would change the weather. That might be a good thing. Modern windmills do not have the conventional vanes and need not be rotated to face the wind.

When the sun is overhead on a clear day one square foot of ground receives .129 kilowatts of heat, or .01 to .04 kilowatts of power, depending on how you dispose of the heat rejected by the process. The simplest way to use the energy of sunlight is to intercept it with thin black wafers of silicon which convert it directly into electric power—as much as 10 watts per square foot. Until recently, these wafers have been so expensive that they could be used only to power satellites but the price is coming down fast now, due to new techniques of manufacture, and this may soon become a practical means of using solar energy. Or you may collect the sun's heat in water, in pan-shaped containers, and use it directly in your bath, or to heat your house. This has been done economically for many years in our South. Or you may heat the water, enclosed in pipes, well above its normal boiling point, and use it to run a steam engine and so produce power. To get the water hot enough for this purpose you have to focus the light of the sun on the pipes using lenses or mirrors, and these have to be rotated to conform to the instantaneous position of the sun as it moves through the sky.[4]

The intermittent power sources described above are attractive aesthetically but they share a fatal defect — the installed cost per kilowatt for firm power (power available twenty-four hours a day) is two to five times as much as the installed cost of firm power sources based on nuclear reactors or fossil fuels. This is a brutal fact of nature which all our cleverness cannot circumvent. Our gimmicks — laboratories, studies, think-tanks, and conferences of experts, are of as little avail in this area as shoveling against the tide. We just have to live with this fact, though we may wish to examine it a little more closely.

196

Suppose that we were to rely, even in part, on intermittent power sources. We have two means of compensating for the times they are out of operation because of drought, clouds, night time, lack of wind, or unfavorable phase of the tide. One is to build standby nuclear or fossil-fueled plants. The cost of these must be added to the cost of the intermittent plant. Or we may make our intermittent plants bigger than they would have to be if they could operate continuously, and provide means to store most of the power they generate when operating and draw upon this store when clouds, darkness, calm, or slack tide shuts us down. The obvious, and probably the best, way to store such large blocks of power is to use any power which may be available after meeting the demands of consumers, to pump water from a lower to a higher level, and then, during periods when the plant is shut down, to use this water to run conventional hydraulic plants. Here, too, the cost of such reservoirs and hydraulic electric generators must be added to the cost of the intermittent generating plant to get a figure comparable with that of a fossil fuel-fired plant which would meet the same consumer demand for power.

The place of water storage and power regeneration need not be near the primary installation. If any plan for wholly natural power sources or a mix of these with nuclear and fossil fuel-fired plants is to work, these must be distributed all over the United States and interconnected in such a way that power can be transferred clear across the country with insignificant losses. It will be sunny in the Mid-west when it is cloudy in Boston, and vice-versa. There are well known stability difficulties in the way of doing this at 60 cycles (the quarter wave length at 60 cycles is only 777 miles as compared with the 3,000 mile width of the country) and probably the main power trunks would carry direct current at several million volts. Conservationists who advocate doing it this way should note that tubular conductors a yard or more in diameter spanning the country hung from mammoth towers will not contribute to our aesthetics. They should also note that if we are to use water storage to carry us through dips in output

197

from the main power plants we shall have to dam many high mountain valleys and convert them into reservoirs for this purpose.

At this point we have accumulated enough facts to reveal the outline of the main problem. No matter how much or how little we spend on the space program during the next few years we have to have more energy, and we shall have to pay for it. The only choices left open to us have to do with the nature of the coin in which we shall pay.

If we go the nuclear route we shall pay in terms of physical danger, real or fancied, to ourselves; in scrapped social concepts, and some disruption as we switch from gasoline powered to electric vehicles. We shall have to give up the pious fiction of the sanctity of human life.

If we go the coal oil shale route we shall have to pay in terms of aesthetics, cutting gashes across our western playgrounds, and putting up with more sulfur dioxide and dust in the atmosphere.

If we go the intermittent power route (power from tides and sunlight) we have to pay in money, putting in installations more costly than those required by the nuclear and fossil fuel options.

These courses of action are not mutually exclusive; any mix of the three is practical. My own preference is to go mainly the nuclear route with enough development of fossil fuel resources so that we can go slow in shifting from gasoline (and diesel) road vehicles and can rapidly become a petroleum exporting nation.

In Chapter 10 we took a look at ourselves to see whether we, as individuals, were adequate to carry out the grand design of evolution. We were not concerned there with the adequacy of government because we were taking a long view, and while the nature of people changes slowly the nature of government changes fast. In this appendix we are looking at a short-term problem — how to meet our needs, and those of the space program, for energy. The physical facts defining this problem have been brought out; we must now examine the political facts which limit what we can and cannot do.

First, the implications of size. We are a bigger nation now than we were, and if we can thread the maze of hazards outlined in Chapter 2 we shall become bigger still. What we can actually do is determined by our size.[5]

Looking further to our economic health we must find a way to insulate ourselves from the effects of crises overseas. We have been badly battered in recent years in economic fights we didn't start; we are punch drunk, and we need a period of social continuity to regain our strength. We are vulnerable to manipulation of monetary systems. We must hasten to pin the value of the dollar to food,[6,7] the only commodity which we can be sure of producing, for a long time to come, more efficiently and in greater amount than anyone else. This can be done by taxing farmers heavily on unused land, or land set to luxury crops, and exempting them partially or entirely from taxation on income from meat production, soy beans, and grain. We must build up stockpiles of these last so that we can meet overseas need for them out of inventory rather than production.

The thing that shakes us most in these modern times is the rising cost of living. This is closely tied to the price of food, and this to the price of beef. If we can rearrange our farm policy so as to insure us all the grain-fattened beef we can eat we are not likely to complain about anything else.

Just to meet our conventional energy needs for the next two decades, let alone those of the space program, will involve such a tremendous dislocation of our customary patterns of life as to threaten the stability of government. If we act without careful thought we shall meet each crisis, as it comes, by setting up new governmental bureaucracies. This way lies instability and disaster. We will have more people regulating us than we can support. The alternative is use of the power to tax to make private industry and organized labor do what has to be done. Perhaps organized labor, which will have less to do than it does now if we can attain some measure of tranquility, can be persuaded to undertake the program of retraining and reshuffling of our work force and the pasturing of those unwilling or unable to adapt to changing conditions.

If we make an all out effort to conserve energy, and specifically to shift commuting and shopping travel from gasoline powered automobiles to public transportation and electric powered automobiles, and if we keep domestic production of petroleum and natural gas the same as it is now, and if we meet our needs for additional oil, as fast as we can, by strip-mining and converting coal and oil-shale, and if we meet our needs for additional electricity partly by building additional oil-fired generating plants but mostly by building nuclear generating plants, our capital investment in these things, and our imports and exports, may be about as follows.

Table II

Projected Oil Imports and Money Requirements

Year	Oil Imports	Oil Exports	Annual Investment
1976	600	0	40
1978	520	0	70
1980	460	0	140
1985	240	0	200
1990	0	230	250
2000	0	1,200	300

In Table II the unit of energy is one million kilowatts and the unit of money one billion dollars. The total investment in new coal oil shale conversion facilities and power plants may be about 2,400 billion dollars.

Notes

1. One gallon of petroleum weighs about eight pounds. One barrel of petroleum contains 46 gallons which weigh about 368 pounds. There is some confusion in petroleum statistics, different authors using anything between 52 and 36 gallons as the content of a barrel.

2. Battery powered cars may also be competive for long distance, high speed runs. You would have to change the bat-

tery — not recharge it — about every twenty miles but this could be done in less than a minute. You simply drive over a pit at the side of the road where your whole battery pallet drops out and is replaced by a pallet of fully charged batteries. A computer would read off the serial number of your car and bill you, and you would not have to see or talk to anyone. The major highways could be lined with so many of these pits that you would never have to wait.

3. While writing this it has occurred to me that it would be better to operate these power stations totally submerged, anchored, like sea mines, about a hundred feet above the thermocline, with sufficient positive buoyancy so that they could be brought to the surface for servicing. Except while they are being serviced, only power would be brought ashore. There are vast areas near the Bahamas and in the Caribbean where such stations could be used.

4. In this appendix we are concerned with sources of energy which we can use to get the space program started. I have therefore not included power from the sun intercepted by synchronous satellites and beamed from them to earth, because we cannot generate power in this way until we have a substantial work force in space. If we could do it now it would be more attractive than any of the power sources described here, in that it would have little effect on the environment and would be available at full strength twenty-four hours a day. The intensity of the beamed power will be about two thirds that of sunlight, and it can be rectified into direct current at more than 50 percent efficiency.

5. There is a halibut concealing himself on the ocean floor in the deep water off Jeffrey's Ledge, ready to make a mad dash after any suitable food which may pass by. He is a dim-witted creature, but he knows how big he is and that is what is important to him. He does not chase after sea worms or short lobsters any more because he knows that although he could catch and eat them, if he does he will starve to death — he will consume more energy chasing them than he can recover by digesting them. He has become too big to use them economically. He will not tackle a bass or an outsized lobster because

if he did he would be cut to ribbons — he is not yet big enough for that. So he waits, conserving energy and that luscious white meat, until something his size comes along.

6. When you play for keeps you play your own game, not the other fellow's. Any ten-year-old playing marbles understands that. We must base our currency on something — say soy beans — which we have more of than anyone else. India, Russia, France, and South Africa have more gold than we do. If we play that game we lose. We had better continue, as we do now, to sell our gold in the present favorable market. I shall not rest easy while there is any of it left in Fort Knox.

7. We can learn once more from the story of Joseph and Pharaoh, and start building up seven-year stockpiles of foods which keep well. These include corn, wheat, rice, and soy beans.

Appendix 4

The Carbon Dioxide Window

The carbon dioxide[1] in the air acts something like a window regulating the temperature of the earth. The more carbon dioxide we put into the air the hotter the earth should get. If it gets too hot the polar ice caps would melt, the sea level would rise about 250 feet, and some of us would have to move to higher ground.[2, 3]

Now for a closer look at the problem. According to R.M. Rotty,[4] between 1958 and 1970 the carbon dioxide of the atmosphere increased by eleven parts per million (by volume). His calculations show that, other things being equal, this would have raised the temperature of the earth about 1° F. between 1958 and 1970. Actually he found that the temperature of the earth decreased by about .2° F. during this period. Something is adrift; apparently other things are not equal.

The amount of carbon dioxide present in the atmosphere at any one time is about the amount added to it from all sources in two years. Thus, if the carbon dioxide concentration depended only on what went into the atmosphere from decay of vegetation and combustion and what was removed from it by the photosynthetic action of plants, we should have an unstable situation, vulnerable to chance fluctuations of weather. Actually the carbon dioxide in the air is in thermodynamic equilibrium with the vastly greater store of carbon dioxide in the ocean. We now have to ask whether, if the earth began to heat up, and the ocean with it, the ocean would tend to put more carbon dioxide into the atmosphere.[5] If this is the case we could have a runaway situation, possibly catastrophic, since the hotter the earth and the oceans got the more carbon dioxide the ocean would pour into the atmosphere, and the more carbon dioxide in the atmosphere the hotter the earth and the oceans would get. Everything depends on whether the ocean tends to give off or absorb carbon dioxide as it gets warmer.

So far as I can find out this temperature effect has not been measured and we have to resort to theory. Ocean water is too complex a material to admit of direct mathematical analysis; we shall have to reason by analogy with the case of pure water in equilibrium with carbon dioxide and solid calcium carbonate (calcite).

The relevant chemical reactions are:

1. $$CO_2 + H_2O = H_2CO_3$$

2. $$H_2CO_3 = H^+ + HCO_3^-$$

3. $$H_2CO_3 = 2 H^+ + CO_3^{--}$$

4. $$CaCO_3 \text{ (calcite)} = Ca^{++} + CO_3^{--}$$

5. $$H_2O = H^+ + OH^-$$

The National Bureau of Standards has tabulated the energy and entropy changes associated with these (and many other) reactions. From these I have computed the equilibrium constants for the above reactions at several temperatures.

Table I

Equilibrium Constants of Chemical Reactions

Reaction	77°	60°	50°	40°
1	3.720×10^{-2}	4.805×10^{-2}	5.631×10^{-2}	6.641×10^{-2}
2	4.253×10^{-7}	2.632×10^{-7}	1.955×10^{-7}	1.435×10^{-7}
3	1.700×10^{-17}	1.497×10^{-17}	1.256×10^{-17}	1.046×10^{-17}
4	4.536×10^{-9}	5.339×10^{-9}	5.905×10^{-9}	6.559×10^{-9}
5	1.000×10^{-14}	8.571×10^{-15}	7.787×10^{-15}	7.047×10^{-15}

There are six variables: unionized carbonic acid, bicarbonate ion, carbonate ion, hydrogen ion, hydroxyl ion, and calcium ion; and the above equilibrium constants provide five relations between them at each temperature. The sixth relationship, which makes solutions possible, is the requirement for electrical neutrality.

The elegant way to proceed at this point would be to duck the solution of these equations, differentiating them with respect to temperature as they stand, since the "constants" (parameters would be a better term but I adhere to the customary usage) are known functions of the temperature. I tried this but the expressions which I got were so long that I could not be certain they were correct. Instead, I solved the simultaneous equations numerically at 60°, 50°, and 40° F. As a reference point I use the solution at 50° F. for a solution saturated with calcium carbonate and in equilibrium with air containing 325 parts per million of carbon dioxide. This is about what we may have early in 1976. Two cases had to be distinguished: the short term case in which the water does not have time to come to equilibrium with solid calcium carbonate, and the other in which such equilibrium is reached. The first case is analogous to that of ocean water out of contact with chalk or coral formations, and the second to that of ocean water in equilibrium with these. Because of the slow circulation of the bottom waters of the ocean the long term case may be applicable only after several hundred years.

The numerical solutions are given in Table II below.

Table II

Numerical Solutions of the Equilibrium Equations

	60° F		50° F	40° F	
	unsat.	sat.	sat.	unsat.	sat.
Equilibrium pressure of carbon dioxide, parts per million of atmospheric	322	285	325	330	368
Concentration of					

unionized carbonic acid, milligram mol equivalents per liter	.01545	.01371	.01830	.02192	.02446
Concentration of bicarbonate ion, milligram mol equivalents per liter	.8661	.8697	.8604	.8533	.8478
Concentration of carbonate ion, milligram mol equivalents per liter	.01049	.01992	.01329	.01687	.01492
Total carbonate concentration milligram mol equivalents per liter	.8920	.8953	.8920	.8920	.8872
Contained carbon dioxide, milligrams per liter	39.25	39.39	39.35	39.25	39.04
Ph	8.328	8.382	8.381	8.433	8.383

As is seen from the top line of Table II, the equilibrium pressure of carbon dioxide is lower the higher the temperature in both the short term and long term cases. We can put all the carbon dioxide we want to into the atmosphere with no danger of changing the temperature of the earth.

Notes

1. Carbon dioxide absorbs radiant heat (light) in the far infrared. It is almost completely transparent to sunlight, and so has very little effect on the amount of heat reaching us from the sun; but the heat radiated by the earth is almost all in the far infrared and so regulated by the carbon dioxide in the air. Thus an increase in the carbon dioxide content of the air should decrease the flow of heat away from the earth and so build up the average temperature of the earth.

2. The first effect of a rise in the temperature of the earth would be to lower the sea level. More snow would fall on the polar ice caps and accumulate there without much

effect on the melting and runoff from these caps. The sea level would not begin to rise until the earth got so hot that the runoff from the polar caps became equal to the snowfall on them.

3. We can duck the carbon dioxide window problem if we want to, at least in so far as the space program is concerned, by using hydrogen instead of hydrocarbons as the primary rocket fuel. This may be what NASA has in mind; the planned shuttle rockets burn no hydrocarbons. Commitment to a no-hydrocarbon policy would more than double the cost of fuel for the shuttle service.

4. As reported at the ASME-IEEE Joint Power Generation Conference, New Orleans, La., September 16-19, 1973. See also *Time* Magazine, Nov. 11, 1974.

5. I am aware that, locally, the carbon dioxide content of the air may not be in thermodynamic equilibrium with the carbon dioxide at the ocean's surface. The carbon dioxide contents of both media fluctuate from time to time and from place to place, and in some places the photosynthetic action of the marine plants removes carbon dioxide faster than it can be restored from the air. Also, the carbon dioxide pressures in equilibrium with ocean water depend on the temperature of the water, which varies from place to place. The equilibrium referred to here is global, not local.

Appendix 5

Design of a Shuttle Service

The present NASA plan for a first generation shuttle comprises an aircraft, an external fuel tank larger than the aircraft, and two solid-fuel boosters. The aircraft fuel is hydrogen plus oxygen. During flight the fuel tank drops off into the sea and is lost. The two booster rockets also drop off into the sea and are recovered, repaired, and used again. Only the aircraft arrives in orbit; from there it flies back to earth, is partly repaired and partly rebuilt, and is used again. I have not seen or heard any official estimate of how much it will cost to lift materials into orbit using this shuttle but my own estimate places this cost at upwards of $100 a pound.

Neither this shuttle nor its logical descendants will serve the purpose. To get on with the space program we need a shuttle which will carry bulk cargo into orbit at less than $5 a pound. We shall also need a few shuttle buses to carry people into orbit at less than $20 a pound.

The main reason why NASA's proposed shuttle service would be so expensive is that it is a manned vehicle. The crew have to be protected from both the cold of the upper atmosphere and the heat of re-entry. They have a pressurized cabin and sanitary facilities. All this adds to the weight. There is really nothing for a crew to do in this vehicle. Initial and final guidance are better supplied from the ground, and docking guidance from the satellite receiving station. In an unmanned ship the temperature of the whole interior can be allowed to drift up to about 400° F. There is no essential part of the ship which cannot operate indefinitely at this temperature, and by eliminating crew cabins and support facilities thousands of vital pounds can be saved.

The NASA aircraft is designed to plunge steeply into the atmosphere on the return to earth. Impact with the atmosphere brings all its exterior surfaces to white heat and parts of them are actually burned off. If you put on an extra pound of weight, say for crew quarters, the plane has to plunge that much steeper, and the surfaces get that much hotter, and you have to add another .9 pounds to take care of the extra mechanical forces and heat. Against that, a pound saved by eliminating some unessential function, such as a lavatory, is pure gold—the plunge is less steep, the mechanical forces less, and you can take .9 pounds off the weight of the structure as no longer needed. As you add weight you go into a vicious downward spiral—literally.

A second reason why the operation of the NASA design shuttle would be so expensive is that it carries the main engine, which is needed only during the first half of the flight, all the way up into orbit and down again. This adds several thousand pounds at the time and place where weight hurts most.

A third major reason why the present NASA design of the shuttle would be too expensive to operate is the NASA choice of fuel. Lift for lift, the hydrogen-oxygen fuel costs about three times as much as a hydrocarbon-oxygen fuel, and the solid fuel which NASA plans to use (a mixture of ammonium perchlorate with powdered aluminum and a plastic binder) costs about ten times as much as its hydrocarbon competitor. Probably the hydrogen-oxygen fuel is the best choice for the second and third stages, where the higher nozzle velocity obtainable with hydrogen is an advantage, but at lift off you want lower nozzle velocities, and a hydrocarbon fuel is definitely indicated.

A fourth major reason why the NASA design is too expensive is that the external fuel tank, the largest single piece of equipment in the complex, can be used only once.

Minor reasons why the NASA design is too expensive include the fact that the solid-fuel rocket boosters have to be recovered at sea, towed in to port, and trucked to the space port, and that the aircraft structure has to be rebuilt, rather than serviced, between flights. Also, since the aircraft is fixed-wing, the aerodynamic drag on it during the first half of the upward flight is excessive.

When any vehicle descends from orbit to earth a large amount of heat is generated. The total amount of this heat depends only on the mass of the body and not at all on its structure or the path of the descent. Unless this heat is dissipated almost as fast as generated the vehicle will be melted down. There are only two ways in which this heat can be dissipated: one is by radiation and one is by conduction by the air (when the air is in motion relative to the vehicle, as it is in all practical cases, this process is called convection). Any practical re-entry vehicle loses some heat by convection and some by radiation, but all re-entry vehicles up to now have lost most of the heat by convection. This inevitably involves scorching and major repairs if the vehicle can be reused at all. Loss of heat by radiation is a slower process but can be accomplished without the vehicle getting very hot or being damaged in any way. To have a low cost shuttle service we have to arrange to get rid of almost all the heat of re-entry by radiation.

This problem was first encountered in connection with intercontinental ballistic missiles. These have to come in fast; there is no time for them to radiate appreciable heat so their surface must get very hot and get rid of its heat by convection. Next came re-entry vehicles carrying people; these likewise had no wings, had to come in fast, were scorched, and were not reused. In these missions this did not matter; the total cost of the missions was so large that loss of a vehicle had no appreciable effect on it. We are now at a point where economy is the overriding consideration; we must come down slowly, and the only way to do this is by fitting wings to the vehicle.

The NASA design has wings, but for another purpose—so that it can land on a conventional airstrip. Any design which has wings adequate to keep it aloft long enough to radiate its heat will be able to land on a conventional airstrip; indeed, it will be able to land at speeds less than sixty miles per hour.

We decide to use wings, and we would like to make them long and thin like those of a glider, but we see at once that wings of that type, if they are big enough and light enough to give us the slow descent that we need, will break off under the aerodynamic stresses of rocket takeoff. We shall therefore have to use folding wings of some sort. Our experience with folding wings has been unfortunate. We are in process of junking a military plane which has them because we have not been able to make them work. Looking closely at our failures to date we find that they occurred because the wings had to be folded and unfolded under load. In the case of the shuttle service proposed here the wings are folded about the body of the aircraft prior to launch, while it is still on the ground, and under no load, and they will be unfolded while the vehicle is in orbit, again not under load. The problem is easier than that of the folding-wing military plane, and quite solvable.

The wings must fold completely around and close to the body to reduce aerodynamic drag on the way up, and unfold after the vehicle reaches orbit. This requirement practically forces us to use arrangements very much like a bird's wings. The main spars will not be lifting surfaces but purely strength members, and are jointed in at least two places. The lifting surfaces are long, thin blades, internally braced, and not necessarily overlapping when the wing is fully extended. There may be as many as fifty of these blades per wing. They will be of unequal length and chord.

What is visualized here, in orbit and before unfolding the wings, is an empty fuel tank, an empty cargo bay, a packet of remotely controlled instruments and actuating devices, and a relatively tiny inboard diesel engine to supply power to a propeller during the approach to the landing field, plus

212

of course the wings wrapped around the tank and cargo bay. Mechanics at the orbiting receiving station will unfold and pin them securely in the flight position, fit an external propeller to the engine shaft, and attach a small reusable rocket which will burn for at most a few seconds to drop the speed of the vehicle below orbital. It will then slowly coast into the atmosphere, sail there long enough to dissipate the re-entry heat, glide to the airport, and land there, remote controlled, like a conventional plane.

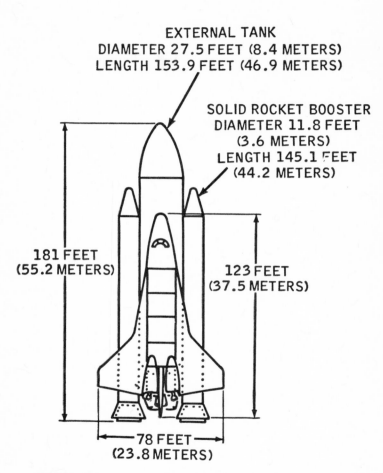

EXTERNAL TANK
DIAMETER 27.5 FEET (8.4 METERS)
LENGTH 153.9 FEET (46.9 METERS)

SOLID ROCKET BOOSTER
DIAMETER 11.8 FEET
(3.6 METERS)
LENGTH 145.1 FEET
(44.2 METERS)

181 FEET
(55.2 METERS)

123 FEET
(37.5 METERS)

78 FEET
(23.8 METERS)

NASA first generation shuttle — February 1974

Appendix 6.

Control of the Weather

It is possible to control the weather by means of operations in space. Briefly, this can be done by placing great numbers of mirrors in orbit and rotating them so as to deflect the heat of the sun toward places on the ground to be heated and away from places to be cooled. Whether or not to do this, and when, are mainly questions of cost.

By launching mirrors in such numbers that their combined area is about one percent of that of the sky, it will be possible to deflect dangerous storms into channels where they will do little harm. The first cost will be about one percent of the total annual tax take at all levels of government—federal, state, and municipal—of all countries, and the annual operating cost will be about five percent of that, including maintenance and replacement of lost or defective units.

By launching enough mirrors to cover about a tenth of the sky we can have completely predictable weather without substantial change of climate. Thus we could know several years in advance that June tenth, say, of the current year will be fine in Boston, though Boston will have about the same proportion of fine and rainy days as it has now. With this number of mirrors, dangerous storms and tornados will be completely eliminated, as will flooding and destruction of crops by wet or drought. The first cost will be about ten percent of the annual tax take of all countries.

By placing enough mirrors to cover about half the sky we could change the climate in every part of the earth, in such a way as to make deserts into gardens without impairing the

215

productivity of areas now in cultivation, in addition to providing predictable weather and immunity to storm, flood, and drought damage. In this, as in the other cases, the launching of the mirrors will be spread over many years. Even the first few will do some good and the effect will be gradual and cumulative, progressing through storm and flood control through reliable long term weather prediction to climatic change. The three phases of the program listed above are arbitrarily chosen to give some idea what can be accomplished at different levels of investment.

The mirrors are not as easy to design, install, and operate as the above might suggest. The system is subject to political and technical as well as financial constraints.

It may prove economically and politically sound to launch the first few mirrors in more or less equatorial orbits so that they will pass over only the southern fringes of Russia[1] and affect primarily our own weather, but before many mirrors are in orbit some of them will have to go near the poles and pass over every spot on earth. Control of the weather is a worldwide venture. What is done to deflect a hurricane away from New England will affect the weather everywhere on earth to some degree, perhaps increasing the rainfall in the Sahara and making the autumn temperatures in Siberia a few thousandths of a degree colder. Any country which can influence global weather has political leverage over every other country, and a nation which can have a substantial influence on worldwide climate actually controls every other nation. Furthermore, the mirrors, besides serving primarily to regulate the weather, are potent weapons. You can burn up a city by focusing on it two or three times the solar heat it normally gets, and if you wish to minimize the casualties of war you can force your will by burning up crops.[2] It would seem that if more than a very few mirrors are launched they must be controlled by an international authority.

The most serious technical limitation on the mirrors is that they must be used in such a way as not to change the average temperature of the earth very much. If we heat the earth the polar ice caps will melt, the sea level will rise, and we shall all

216

have to move inland.³ If we cool the earth the ice caps will grow and we shall provoke another ice age. This limitation hurts because, using the mirrors, it is easier to heat the earth than to cool it. You can focus sunlight on a place using mirrors anywhere off to one side, but you can cool that same place only by passing mirrors directly between it and the sun. Thus the mirrors will have to be used most of the time as sun-shades to compensate for the few times they are used as burn-ing glasses.

A second technical limitation derives from the force exerted on the mirrors by the sunlight they deflect. This force pushes against familiar terrestrial objects as well, but these are so massive that the effect goes unnoticed. The mirrors will have to be extremely light, fragile objects in order to keep the cost within the bounds of practicality, and because they are so lightweight the force of sunlight on them may be as much as one percent that of gravity. Thus, in addition to orbiting, they will sail, much as a yacht does (the most sail a yacht can carry in a strong wind exerts a force less than ten percent that of gravity on it). As the mirrors function they will be continually changing their orbits. To prevent them from colliding, or dip-ping into the atmosphere and burning up, or wandering off into orbit about the sun or moon, they will have to receive from the ground special sailing instructions. Since the angle they make with the rays of the sun will have to be determined in part by the necessity of keeping them in compatible orbits, their effectiveness in controlling the weather will be corres-pondingly reduced. It may be too that blasts of gas from the sun—the so called solar wind—will complicate the sailing problem. My calculations indicate that the force of the solar wind will ordinarily be much less than that of sunlight, but that on rare occasions the solar wind may push harder. The trouble, if any, comes from the fact that while the force of sun-light is completely predictable, and can be calculated and compensated for well in advance, one gets only a few minutes warning of a strong solar wind. The computers governing the flight of the mirrors will have to react quickly when this occurs, and a few mirrors may be lost, blown into collisions

217

with others, when a solar storm comes up suddenly.

Tidal forces place a physical limitation on the design of the mirrors. It is ordinarily and correctly assumed that gravitational forces generate no load on a body in orbit, and this is true in the limit of small structures, but in large structures tidal forces may become significant, especially in very fragile things like these mirrors. A one percent increase in the dimensions of a structure increases the tidal forces by about three percent. In the case of these mirrors the tidal forces will be due almost entirely to the pull of the earth, the contributions from the sun and moon being quite negligible in comparison. We should like to make our mirrors as large as possible so that, for a given percentage coverage of the sky, there would be fewer of them to control, and tidal forces may set a limit to the size we can make them.

To see how the mirrors can control the weather we need a physical model of the system which produces it. The polar regions of the earth receive little heat from the sun because its rays strike obliquely there, and because the surface (mostly hoarfrost and ice, not snow) reflects a larger proportion of the sunlight back into space than does the bare ground or foliage of the temperate and tropical zones; but the polar areas radiate heat into space about as well as the other zones do. The result is that the polar regions lose more heat by radiation than they gain from sunlight, and they would become even colder than they are except that the very fact of being cold attracts air to them. This air warms the polar regions and itself becomes cold and dense. The weight of air so accumulated at the poles increases with time, and in about a week becomes too great to be balanced by the momentum of the incoming air. Then, at an uncertain time and place on the rim of a polar region, part of the cold air spills over into the temperate zones and produces our local weather. The important thing is that when the mass of the cold air has built up to the point where it is ready to burst out of the Arctic, the slightest perturbation will cause it to go in one direction rather than another, and at one time rather than another. Thus by cooling a selected portion of the rim of a polar region and heating the rest of the rim slightly, the cold air can be made to spill over at a time and in a sector of our own choosing. So, in a given week, we can choose whether we get the good weather and the Russians the bad, or vice versa, instead of leaving this to chance as we do now.

The thought is not to drive the weather—the sun does that —but to steer it, and steering requires much less power than driving. The power used by a man to control an automobile is less than a thousandth of the power developed by the engine.

Another way to see this is to think of water dripping into a saucer. Eventually the water will fill the saucer and spill over but, no matter how perfectly the saucer may be formed, the water will not spill out uniformly over the rim but rather at some point on it—what point is a matter of chance. The time of spilling is uncertain by at least ten drops. Now suppose that we are able to tip the saucer just a little and at a time and in a direction of our own choosing. The water will spill when and where we please.

The main obstacle to installing a weather control system in space is cost, and the design of the mirrors must minimize cost. Cost is determined principally by weight; we must therefore minimize weight. Weight is determined primarily by the thickness of the metal foil used in building the mirrors. If we make this foil too thick it will weigh too much; if we make it too thin too much sunlight will pass through it rather than be reflected, and the light which passes through does us no good. The best compromise may be a foil which passes half the light through.

This is an old problem. Gold foil is used for placing lettering on glass doors and even on the dome of the State House. Gold is expensive and for hundreds of years now people have experimented with it to find the most economical thickness to use. You can see through present day gold foil; the light which comes through is green, but enough is reflected to give the appearance of solid gold. The optimum thickness for reflecting foils in space is not known, and both data and theory with respect to the optical reflection and transmission of thin metallic films are scanty. My own rough calculations suggest that the optimum thickness may be one half wave length in the metal of light in the near infrared. Light of a particular color has associated with it a characteristic distance, known as the wavelength, ranging from sixteen millionths of an inch in violet light to thirty-two millionths of an inch for red light. Light and radiant heat are the same thing except that the eye can see light, whose characteristic distance ranges between the two above numbers, and cannot see heat, for which the characteristic distances are larger.

The metals or alloys used to make the films have to be chosen from those sufficiently abundant. I have tentatively ruled out boron as too dificult to roll into films; lithium and beryllium as too scarce. Sodium and potassium are too weak mechanically and melt too low to be useful in pure form, but they may be of service as alloying constituents. Magnesium seems to be the best bet; it is presently extracted from the ocean which contains several hundred times as much as would be needed to cover the sky completely. Calcium is also a possibility; the quantity available from limestone, chalk, and coral formations is many times that which could possibly be needed. Aluminum, equally abundant but harder to extract, is also a possibility. Iron may be useful as an alloying constituent. Titanium is a distinct possibility. Chromium and manganese, from nodules on the sea bottom, may be useful in small amounts to make the alloy more opaque. Zinc may also be useful as an alloying constituent. This about exhausts the list of possible metals. In what follows I assume that pure magnesium will be used.

The optimum thickness of magnesium foil may turn out to be about four millionths of an inch. A square of this foil a mile on an edge would weigh one thousand pounds. Enough to cover the entire sky at an altitude of four hundred miles would weigh one hundred twenty million tons. To this must be added an equal weight for girders to stretch the film and optical and electronic equipment, making a total of two hundred forty million tons.

The mirrors could be flat circular discs stretched on a network of tiny girders made up of hairlike filaments of some material, possibly a nonmetal, especially strong per unit weight. For purposes to be explained later there will be, near the edges of the disc, three smaller discs, hung in gimbals, symmetrically placed, so that they can be swung to any orientation with respect to the main disc, and with one face reflecting and the other blackened. At regular intervals all over the disc and in a mast structure there will be triads[4] of rotors with axes at right angles, less than a tenth of an inch in diameter and capable of top speeds of several hundred thousand revolutions per minute, electrically driven, and capable of reversing their direction of rotation. In mast structures, one to either side of the disc, there will be pairs of telescopes, one directed by the mirror computer to point always at the center of the sun and the other tracing the horizon of the earth in a continuous scan.

In operation the discs will be under continuous observation and in communication with radio stations spaced almost evenly over the surface of the earth, which will observe the position and orientation of each disc and give commands. Since there may come to be more than a thousand billion discs flying at the same time, keeping track of them will be a formidable problem, but it is within the capability of present computer technology. Observation may be by means of large dishes like the one at Jodrell Bank, or by interferometric systems. A specific mirror will be challenged by its serial number. It will first give a sharp pulse, which will provide an accurate determination of its distance from the challenging station. It will then give its orientation with respect to sun and earth, as observed by its own telescopes, its rate of rotation about three axes, the speed and direction of spin of its rotors, and the distance and relative course and speed with respect to its nearest neighbors, identified by serial number. This would end the transmission from the disc, which will then receive its instructions to cover the next two hours. The whole transmission will take less than a ten-thousandth of a second. As many as ten thousand big dishes may ultimately be required.[5] The control operation seems like a lot of information to manage but it is less per hour than our own internal computers handle in digesting a single meal, taking the proteins of the food apart and rearranging them as required to maintain our different organs.

The on-board computers will generate almost all of the sailing instructions. Maneuvers to stay in the required orbit and avoid collisions will take precedence over maneuvers to control the weather, but ordinarily this will not require cancellation of a weather control maneuver but merely a slight modification of it. In the aggregate, on-board computers will process more information than those on the ground.

The discs will normally sail with an edge toward the sun; in this position the sun will have the least effect on the disc's orbit, and the disc will have the least effect on the weather. When the time comes to rotate the disc to a different position, or to sail it or to deflect sunlight toward or away from selected points on the earth, the speed of the rotors distributed through the disc structure will be changed. If you wish the disc to go round to the right you speed up the rotors to the left.[6] The problem is the same as that already described of keeping a space station facing toward the sun. To handle the weather problem one may need to be able to turn the disc completely over in as little as one minute. The size of the rotors and the strength of the structure of the disc have to be designed to provide and to stand the necessary forces.

Theoretically, in the first approximation, if a disc started its voyage with all rotors stationary, it would never later be necessary for any of them to exceed a certain rate of speed determined by the highest rate at which the disc has to turn about an axis in its own plane to accomplish its mission, and this speed would be set in the design stage and never exceeded in practice. Actually, things would not work out that way. There would be a slow drift in the rotor speeds even when the disc was sailing with an edge to the sun, and these would accumulate to the point where some of the rotors would fly apart when the disc was maneuvered. It is to take care of this problem that the small discs with one side bright and one side black are included in the design. The pressure of light on these can rotate the disc—very slowly, it is true, perhaps as slowly as one revolution per day—independently of the action of the rotors. By means of these small discs, if a rotor gets going too fast, part of the load is taken off it by using sunlight to rotate the disc in the same direction. The on-board computer senses excessive rotor speeds well in advance of catastrophe and rotates the bright and dark vanes in such a way as to reduce them. Ideally all rotors should be stationary when the disc is flying with an edge toward the sun and the combination of on-board computer and bright-dark discs will be able to maintain this condition quite accurately.

The disc assemblies are so fragile that they can be put together only in the vacuum of space. If an attempt were made to assemble them in air or in the atmosphere of a space station the slightest breeze—as little as might be caused by the motion of a man's hand—would tear the film apart. Construction on forming jigs (frames) weighing tens of thousands as much as the completed discs will have to be completely automated or at most aided by a human operator watching from a distance through a window and telescope. When the disc assembly is completed it will simply be pushed off the jig into the sunlight when it will sail into its appointed orbit. Rolling the magnesium into the sheets of the required thickness will also be done in the vacuum of space.

The cost of moving one pound of magnesium into orbit is estimated at two dollars and twenty-two cents a pound (appendix 1.) The cost of magnesium of adequate purity delivered at a space port is estimated at fifty cents a pound, for a total cost of mirrors of two dollars and seventy-two cents a pound.[7] Thus the cost of the mirrors having a total area the same as that of the sky four hundred miles up comes out six hundred fifty billion dollars.

In estimating these costs, nothing is charged for the labor of space people. They work for subsistence and to pay off the mortgage on the facilities they use (about twenty-five thousand dollars each) and to pay for the schooling of their descendants and for their own retirement. This may seem startlingly different from the way we live now, shuttling money through banks and tax collectors, but the net result is much the same.

Launching the first mirror presupposes the existence of an active space shuttle service, barracks and factories in space, and some specialized machinery; it cannot come sooner than ten years from now. The first units launched will probably be quite small—perhaps less than fifty feet in diameter—because, weight for weight, small mirrors will be more effective in burning up intercontinental missiles than large ones. The first mirrors will probably cost more than ten times as much per unit area as the figure given above; the price will perhaps not get down to that figure until enough mirrors have been launched to cover one percent of the sky. Experience with the first mirrors will lead to changes of design of the later ones; mirrors of quite different shape will work as well together as mirrors of the same shape. The last mirrors built will probably be about a mile in diameter.

I disclaim originality in the design of this weather control system. The problem was thought through by Hermann Oberth and published in a paper entitled "The Rocket into Interplanetary Space" in 1923. I have used some of the technical information and experience which has accumulated since then to elaborate his ideas.

I am not so naive as to suppose that the mirrors actually launched will look much like those described here; I do know that any changes which later, more knowledgeable designers make in these plans will reduce the cost of the system and make it work better. I have worked through the design in detail to make sure that no insurmountable problem has been overlooked and to get an estimate of the cost of the system.

Notes

1. It will be necessary to set up the mirror orbits so as to leave corridors, similar in function to those of a modern airport, through which ascending and descending shuttle traffic can pass.

2. It is tempting to assert that a fleet of mirrors is an effective countermeasure against intercontinental missiles. Enough mirrors to cover half the sky would be a complete barrier to missiles from Russia to the United States or vice versa except during winter nights. The mirrors would be turned to focus the sunlight on the incoming missiles and burn them up. Salvo fire would not avail to saturate the defense; the targets are hundreds of miles apart and the launching sites are also hundreds of miles apart, and it would be impossible to bunch the missiles in time or space in such a way as to get through. But there are those winter nights. To protect us then we would need mirrors in orbit so high that they would be in sunlight even at midnight of the arctic winter, and these would be less effective in regulating the weather than mirrors lower down. Furthermore, such sun-lit mirrors would be further away from the missiles they have to burn up, and this would require that we put up a lot of them to do the job. An adequate all-year defense against intercontinental missiles by mirrors would probably cost more than ten times as much as a complete weather and climate regulating system.

3. Actually, the first effect of raising the temperature of the earth would be to decrease the sea level. For details on this see appendix 4.

The total heat produced, worldwide, by the burning of coal and oil in power plants, domestic heating, and vehicles is now only enough to raise the average temperature of the earth .007 degrees F.

4. It may seem at first that only pairs and not triads of rotors would be needed since rotation of the disc about the axis at right angles to its surface would not affect its performance in controlling the weather. However, rotation about this axis at speeds in excess, say, of one revolution per hour would complicate the control problem, partly because of gyroscopic effects, and partly because such rotation would make it more difficult to keep one telescope aimed at the center of the sun while another scanned the visible horizon of the earth. Such rotation would also complicate the problem of navigation incident to sailing. It seems best to design, as just

described, to keep rotation about the vertical axis slow.

5. Possibly a control system based on lasers would be cheaper than the radar-based system here proposed. A laser-based system would have the immense advantage of providing accurate azimuths and altitudes of the individual discs as well as distances, but there are two serious problems with it. One is that laser technology is not yet sufficiently advanced to do the job. The other is interference by clouds; a disc might have to be designed to sail as much as a week without instructions. This could be done at the cost of some impairment of weather controlling capability.

6. Note that in this context the speed of the rotors may be plus or minus, say plus for clockwise rotation and minus for counterclockwise rotation, considered with respect to designated ends of the axes. To rotate a disc from one position to another it may be necessary to decrease the absolute speed of rotation of some of the rotors.

7. Here, as elsewhere in this book, in estimating costs, only goods and services by people on earth to people in orbit are charged. It is assumed that earth people run the shuttle. Cost estimates are not for the first ten units, or the first million units, but for the last unit of its kind manufactured. Capital charges and research and development costs are assumed to have been written off.